A Manual of Setting-Out Procedures

GALLIFORD & SONS LIMITED
CIVIL ENGINEERING CONTRACTORS

URWICK ORR & PARTNERS LIMITED
MANAGEMENT CONSULTANTS

 CONSTRUCTION INDUSTRY RESEARCH AND INFORMATION ASSOCIATION

SBN 901208 87 6

First published December 1973
Reprinted 1974

Published by
Construction Industry Research and Information Association
6 Storey's Gate, Westminster, London SW1P 3AU

Printed by
Coalville Times Limited, Coalville, Leicester LE6 2QP

PRICE £2·00 Post Free within U.K.

Contents

	Page
INTRODUCTION	ix

PART I
SETTING OUT EQUIPMENT AND GENERAL PROCEDURES

I.	INSTRUMENTS AND MEASURING TAPES	1
	A. Instruments	1
	B. Measuring Tapes	2
II.	OTHER SETTING OUT EQUIPMENT	2
	A. Pegs	3
	B. Ranging Rods	3
	C. Profiles	3
	D. Travellers	4
III.	TEMPORARY BENCH MARKS (TBM's)	8
IV.	RECORDING DOCUMENTS	10
	A. Level Books	11
	B. Sewer Information Sheet	11
	C. General Purpose Information Sheet	13
V.	HINTS ON SETTING OUT	14
	A. General Procedures	14
	B. Use of Sub-Contractors	14
	C. Use of Measuring Tapes	15
	D. Setting Out Angles	15
	E. Setting Out Curves of 30m Radius and Under ...	16
VI.	MODERN INSTRUMENTATION AND TECHNIQUES ...	17
	A. Electromagnetic Distance Measuring (E.D.M.) Instruments	17
	B. Direct Reading Optical (D.R.O.) Instruments ...	18
	C. Lasers	18

PART 2
PIPE LINES AND SEWERS

I.	INITIAL CHECKS	19

		Page
A. Distance Between Adjacent Sewers		19
B. Interference By Drain Connections		19
C. Crossing Existing Services And Sewers		19
D. Manhole Positions		19
E. Manhole Clearance		19
F. Services		20
G. Discharge Levels		20
H. Gully Connections		20
II. SETTING OUT PROCEDURE — LINE AND LEVEL ...		20
A. Centre Line		20
B. Off Set Pegs		20
C. Profiles And Travellers		21
III. MANHOLES		22
IV. JUNCTIONS		22
A. Marking The Position Before Laying		22
B. Marking The Position Before Backfilling		24
V. BACKDROPS		24
VI. DUAL SEWERS IN DUAL TRENCHES		24
VII. CONCRETE BED		25
VIII. DEEP SEWER TRENCHES		25

PART 3
ROADWORKS

I. INTRODUCTION		27
II. GENERAL PROCEDURES		27
A. Proving The Survey		27
B. Dealing With Discrepancies		29
C. National Grid And Its Application		30
D. Accuracy		33
E. Keeping Site Staff Informed		34
III. HOUSING AND INDUSTRIAL ESTATE ROADS ...		34
A. Initial Checks		34
B. Standard Setting Out Procedure — Line		35
C. Standard Setting Out Procedure — Level		36
D. Footways		39

		Page
	E. Cambers	40
	F. Kerb Levels	42
	G. Channel Levels With Valleys And Summits	42
IV.	MAIN HIGHWAYS	43
	A. General	43
	B. Establishing The Centre Line	43
	C. Establishing Existing Ground Levels	44
	D. Widening Existing Highways	44
	E. Cuttings	45
	F. Embankments	46
	G. Land Drains	48
	H. Spoil Heaps	48
V.	SETTING OUT CALCULATIONS	48
	A. Setting Out Using Co-ordinates	51
	B. Vertical Curves	56
	C. Horizontal Curves	58
	D. Spiral Transition Curves	61

PART 4
STRUCTURES

I.	INTRODUCTION	65
II.	LOCATION OF STRUCTURES ON THE SITE	65
	A. General Principles	65
	B. Initial Checks	66
	C. Relating One Structure To Another On Site	67
III.	SETTING OUT AND CONTROL OF THE ELEMENTS OF THE STRUCTURE	68
	A. Excavation	68
	B. Piling	70
	C. Stanchion Bases	72
	D. Reinforced Concrete Columns And Walls	72
IV.	MULTI-STOREY STRUCTURES	73
	A. Verticality	73
	B. Height And Level	74
V.	CURVED STRUCTURES	74
VI.	CIRCULAR STRUCTURES	75

Preface

Guidance on site practice for the young engineer is seldom included in civil engineering text books. One reason for this is the difficulty in commissioning practising engineers, skilled in this field, to write such material for commercial publication, because of the cost.

CIRIA members are concerned at the absence of information on this subject and the Association has therefore decided to undertake the publication of guides to site practice, of which this is the first.

The Association is grateful to Urwick Orr and Partners and Galliford and Sons Ltd., for bearing the cost of the preparation and printing of this manual.

Acknowledgement is also due to Mr. J. T. Cieslewicz of JTC Surveys Ltd for his invaluable help and advice.

Introduction

The additional costs and time that inevitably arise on construction contracts through faulty setting out are well known to be considerable. Most setting out errors are caused by two factors —failure to work well established procedures and the use of shoddy equipment and materials.

The purpose of this handbook is to recommend in some detail the procedures and equipment which should be adopted for the setting out of engineering works. It is hoped that the acceptance and use of the recommended procedures will result in better communication and understanding between consulting engineers, resident engineers and clerks of works and contractors' engineering staff and foremen.

All site engineers and others concerned with setting out must understand clearly the responsibilities of both the engineer or architect and the contractor. These responsibilities are normally laid down in one or other of the following Conditions of Contract.

I.C.E. Standard Conditions

"The Contractor shall be responsible for the true and proper setting out of the works and for the correctness of the position, levels and dimensions and alignment of all parts of the works and for the provision of all necessary instruments, appliances and labour in connection therewith.

If at any time during the progress of the Works any error shall appear or arise in the position, levels, dimensions or alignment of any part of the Works, the Contractor on being required so to do by the Engineer shall at his own expense rectify such error to the satisfaction of the Engineer unless such error is based on incorrect data supplied in writing by the Engineer or the Engineer's representative in which case the expense of rectifying the same shall be borne by the Employer.

The checking of any setting out or of any line or level by the Engineer or the Engineer's representatives shall not in any way relieve the contractor of his responsibility for the correctness thereof and the contractor shall carefully protect and preserve all bench-marks, sight rails, pegs and other things used in setting out the Works."

R.I.B.A.

Clause 1 (1) The Contractor shall upon and subject to these conditions carry out and complete the Works shown upon the Contract Drawings and described by or referred to in the Contract Bills and in Conditions in every respect.

(2) If the contractor shall find any discrepancy in or divergence between the Contract Drawing and/or the Contract Bills he shall immediately give to the Architect a written notice specifying the discrepancy or divergence and the Architect shall issue instructions in regard thereto.

Clause 5 The Architect shall determine any levels which may be required for the execution of the Works and shall furnish to the Contractor by way of accurately dimensioned drawings such information as shall enable the Contractor to set out the Works at ground level. Unless the Architect shall otherwise instruct, in which case the contract sum shall be adjusted accordingly, the Contractor shall be responsible for and shall entirely at his own cost amend any errors arising from his own inaccurate setting out.

Both these conditions place the onus of responsibility for setting out the works on the contractor. However, site engineers must never forget that many people besides themselves will require to make use of their initial and detailed setting out of a job. Hence one of the site engineer's most important duties is to make certain that all others on site can understand what he has done. The adoption of the procedures and colour codings recommended in this manual will reduce expensive errors which arise from misunderstanding of unfamiliar conventions at site level.

The best results will be achieved if the engineer and contractor agree at the start of the contract to use the recommended procedures. At least, the contractor should notify the engineer in writing of the procedures and colour codings it is proposed to use.

All the procedures and equipment described in this manual are based on practical experience, and have been proven over a number of years. They are applicable to the majority of construction contracts.

There is, however, a growing tendency to use more advanced survey techniques, particularly on major civil engineering contracts in urban areas where accuracy of setting out is vital.

A description of the more important techniques, together with some worked examples, is contained in Part 3, Section IIC and Section VA. A brief description of the more modern instrumentation is given in Part I, Section VI.

For the sake of convenience, the title "site engineer" has been used throughout, but it is intended to apply to all persons who are concerned with the actual process of setting out.

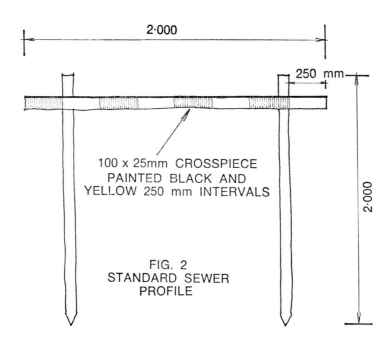

FIG. 2
STANDARD SEWER PROFILE

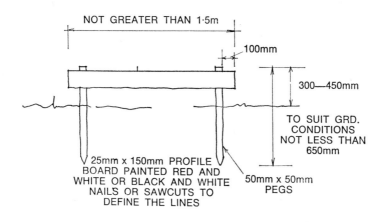

FIG. 3
STANDARD CORNER PROFILE

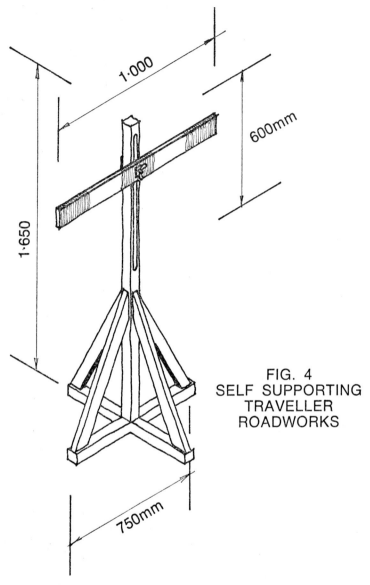

FIG. 4
SELF SUPPORTING
TRAVELLER
ROADWORKS

2. *For Sewers*

The recommended traveller for sewer work is shown in Fig. 5. The cross-piece should be painted alternately yellow and black. Travellers should always be made on site under the supervision of the site engineer.

Traveller lengths should be computed from the *invert level of the pipes*, i.e. travellers should be made the pipe wall thickness (x mm) longer than the distance from the top of the cross-piece to the invert level of the pipe.

A stout 200mm steel shelf bracket should be screwed to the

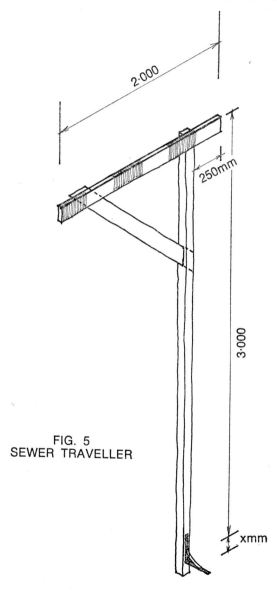

FIG. 5
SEWER TRAVELLER

traveller with the bottom of the bracket fixed at invert levels.

Where a concrete bed is to be provided in a trench, the traveller length should still be based upon pipe invert level. A wooden extension piece should be screwed to the traveller base extending the length of the traveller by a dimension equal to the thickness of the concrete, as shown in Fig. 6.

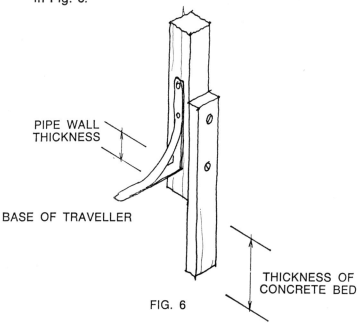

FIG. 6

III. TEMPORARY BENCH MARKS (TBM's)

The work involved in transferring a level from an Ordnance Bench Mark to a temporary bench mark on site is expensive in time and money.

The position of temporary bench marks should be carefully planned in good time and take account of overall contract arrangements. Care must be taken to ensure that progress of the works will not make a TBM useless at a later stage of the contract.

Permanent features should, whenever possible, be used as positions for TBM's. Existing walls, plinths, door steps, etc., can be selected using steel bolts grouted into the existing structure.

Where no suitable permanent features are available, level stakes must be used. Stakes should be made from 50mm x 50mm angle iron 600mm long or from a similar length of reinforcing bar or dowel pin. The stakes should be driven into a previously dug hole down to the determined level as shown in Fig. 7.

Concrete is placed round the stake to approximately 12mm from the top. The reference number of the TBM is then marked in the wet concrete and the top of the stake painted blue.

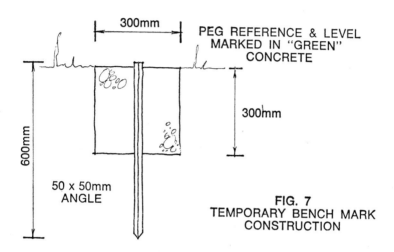

FIG. 7
TEMPORARY BENCH MARK CONSTRUCTION

TBM's and other important reference points must be protected if there is any danger of them being disturbed, as shown in Fig. 8.

If cement and aggregates are not available on site, use bags of dry pre-mixed materials, adding water at the time of use.

Temporary bench marks must never be more than 100m apart on any site and should be set up so that any closing error does not exceed 5mm. The first TBM in a scheme should be established from an Ordnance Bench Mark, agreed with the engineer in writing. It is particularly important to agree which Ordnance Bench Marks are to be used as datum points.

Discrepancies between Ordnance Bench Marks do occur. Those established on old structures can be suspect and the site engineer must check with the Ordnance Survey Department for the latest values.

If working from a previously established TBM, the site engineer should always verify the reduced level personally by levelling from an agreed Ordnance Bench Mark.

If a scheme has been designed using an assumed datum, and its position and reduced level is shown on the contract drawings,

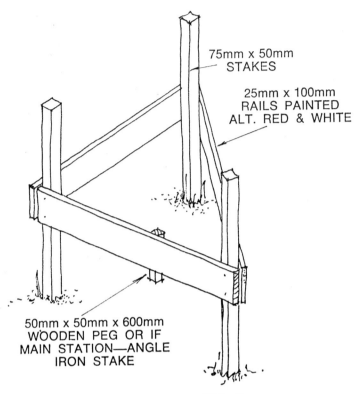

FIG. 8
PROTECTION OF IMPORTANT
REFERENCE POINTS

site engineers should always confirm with the designer, *in writing,* that it is in order for the datum to be used.

The position of all TBM's must be marked on the site plan with their reduced levels and reference numbers. It is also advisable to display a sketch and list of all TBM's in the site offices. The list should indicate the reference number, reduced level and location of each TBM. Finally, site engineers should make a point of checking TBM's at regular intervals. It is good practice to transfer levels from stakes to permanent positions on the works when they are suitably advanced.

IV. RECORDING DOCUMENTS

It is essential that information produced by the site engineer for and during setting out operations is recorded clearly, so that it can be later understood by others, if necessary.

The best practice is to use standard pro formas in the form of site information sheets. These sheets can be printed in pads and sufficient copies made for distribution to all concerned. One copy of each information sheet should be retained by the site engineer.

A. Level Books

Site engineers should be supplied with standard level books ruled for the "Collimation" or height of instrument method for computing reduced levels. Level books are part of the record of works in progress and may be used for valuation purposes. They must therefore be kept in a clean and tidy state so that others can understand the descriptions and computations recorded.

B. Sewer Information Sheet

The purpose of this sheet is to ensure that general foremen and gangers concerned with the management of the work are fully acquainted with what is required. The sheets also form a record for future valuation purposes. To assist the site engineer when completing the information sheet, a check list of points to be considered should be printed in the inside cover of the pad. The list should include the following:
1. Pipe diameter
2. Class of pipe
3. Type of sewer pipe, e.g. concrete, flexible joints, salt-glazed, P.V.C. etc.
4. Manhole number and type with internal dimensions
5. Invert levels
6. Details of offset pegs, e.g. distance from centre of sewer, reduced level of peg, reference no. of peg.
7. Position of junctions and the angles they are to be set to
8. Lengths of concrete bed, bed and haunch and surround
9. Lengths of granular bed and fill
10. Length of sewer between manhole centres
11. Approximate depth of sewer at each profile board
12. Length of travellers
13. Position of profile boards and height of top edge of board above off-set peg.
14. Position and details of back-drops
15. Positions of existing services crossing the sewer—mark the positions in red
16. Direction of flow

The sewer information sheet is completed by sketching the run of sewer between manholes to the right or left hand side of the sheet depending on which side of the sewer the off-set peg is sited, looking up the flow. All the relevant information on the check list is included.

An example of a completed sheet is shown in Fig. 9.

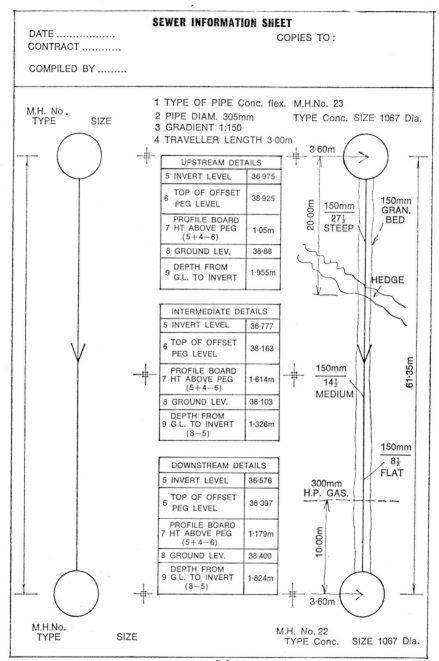

FIG 9

EXPLANATORY NOTES TO FIG. 9

1. Details are filled in on either diagram depending on which handing the Off-Set Pegs are placed
2. 150mm—Dia of junction.
   $27\frac{1}{2}$—No. of pipes from M.H. Wall.
   Steep—Angle of junction from horizontal.
3. Show brick manholes by drawing a square in the circle giving internal dimensions and wall thickness.
4. All dimensions on the sewer taken looking up-stream.

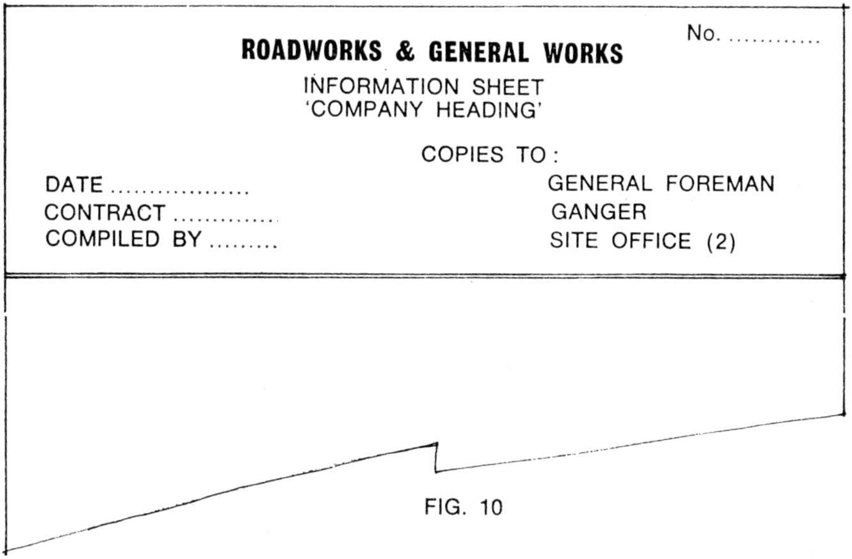

ROADWORKS & GENERAL WORKS   No. ....

INFORMATION SHEET
'COMPANY HEADING'

COPIES TO:

DATE ....   GENERAL FOREMAN
CONTRACT ....   GANGER
COMPILED BY ....   SITE OFFICE (2)

FIG. 10

C. General Purpose Information Sheet

For roadworks and other works a general purpose information sheet should be used to record information and instructions for general foremen and gangers.

Sketches should be used as often as possible to describe what is required. Levels and accurate measurements should always be recorded together with distances from any fixed reference point.

A simple type of general purpose information sheet is illustrated in Fig. 10.

V. HINTS ON SETTING OUT

A. General Procedures

Before setting out any section of the works, site engineers must carry out a preliminary inspection of the site and note any obstacles or features, for example, site hutting, which are likely to interfere with the setting out or with the actual works. These must be reported to the agent or resident engineer immediately so that any necessary action can be taken in good time.

Site engineers must check all drawings and contract documents carefully against the actual site conditions and report any discrepancies in writing. Failure to spot any errors in the drawings may well lead to extra cost at a later stage — and become the subject of a claim.

Before setting out, verify the levels of any temporary bench marks, checking the master TBM against an Ordnance Bench Mark as indicated in Section III.

Site engineers must make a habit of inspecting the site daily, watching for any signs of disturbed profiles or pegs. If in doubt, check the levels and measurements again. In particular, the reduced level of TBM's and the position of location pegs should be checked at regular intervals.

All level books and other setting out documents should be carefully filed and retained at least until the end of the contract. Each document must be clearly headed with the site engineer's name and must indicate to which section of the works it refers.

Finally, any errors in setting out must be reported as soon as they are discovered. Nothing is gained by concealing errors which must eventually come to light, and early action to correct errors will save money in the long run.

B. Use of Sub-Contractors

The responsibility for setting out the works rests with the contractor. Where sub-contractors are employed, the site engineer must familiarise himself with the form of the sub-contract being used. The sub-contractor will usually set out his own work in sufficient detail for his labourer to proceed with the work. This may or may not coincide with the contractor's limits of accuracy of line and level. The site engineer should therefore check line and level of the sub-contractor's work at frequent intervals, preferably with the sub-contractor's representative.

Specialist sub-contractors are often used for earthmoving, thrust borings or the driving of headings. If line or level in these operations goes beyond the required limits of accuracy, rectification is normally very costly.

In addition to periodic checks on line and level, the site engineer should note the class of subsoil that is being encountered and, in particular, any unforeseen obstacles in the path of the excavation. This information may well be important in the event of any claim.

C. Use of Measuring Tapes

The following points must be remembered when using measuring tapes:
1. Always measure at the same level and keep the tape as near the ground as possible to avoid bellying in the wind.
2. Pull the tape sufficiently to remove any kinks — but do not over-stretch it.
3. Make sure that the tape is not obstructed by stones, tree roots etc. between the pegs.
4. Keep the tape clean so that the figures can be easily read. Check the tape regularly against a known length.
5. Use offsets to measure round or over obstructions.
6. Use cross-checks wherever possible i.e. an overall length with the sum of intermediate lengths.

D. Setting Out Angles

Simple geometrical rules can often be used to set out angles, but in cases where a high degree of accuracy is required a theodolite should be used.

The following diagrams illustrate the more common methods:

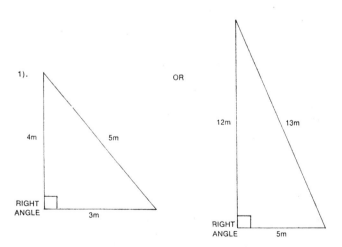

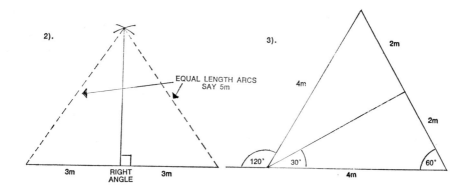

E. Setting Out Curves of 30m Radius and Under

1. Circular curves can be set out by using a steel tape from the centre point, placing pegs equi-distant round the circumference of the circle about 15° of arc apart or at 10m intervals.

   If the centre point is inaccessible, the curve may be set out by offsets along the long chord. The method is shown in Fig. 11.

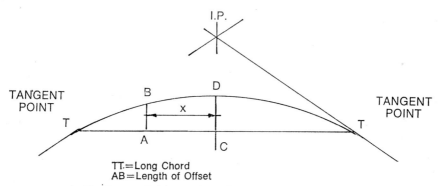

TT = Long Chord
AB = Length of Offset

1. The Tangent Points are determined from the intersection point
2. A Builder's Line is strung along line of chord TT
3. The length of right angle offsets is determined from the formula:—

$$AB = \sqrt{R^2 - x^2} - \sqrt{R^2 - \left(\frac{L}{2}\right)^2}$$

Where
R = Radius of Curve
x = Distance from 'C' the Chord Centre
L = Length of Chord TT

FIG. 11 SETTING OUT A CIRCULAR CURVE USING OFFSETS FROM ITS CHORD.

VI. MODERN INSTRUMENTATION AND TECHNIQUES

The need for complex structures like the Gravelly Hill Interchange has resulted in the increased use of instruments which are capable of a high degree of accuracy coupled with speed of reading. These instruments are expensive to buy when compared with the cost of more traditional instruments. In order to maximise the return on the money invested in such instruments, they must be in almost continuous use.

There are a number of specialist firms which will undertake to supply the services of a trained engineer/surveyor together with the appropriate instrument for setting out work of a more specialist nature. Unless a construction company has sufficient work and trained surveyors, it is likely that use will be made of such specialist firms. But, in any event, a site engineer should be aware of the capability and nature of the more modern instruments, so that he can advise contract management to take the appropriate action when he feels that the solution to a particular setting out problem will be best answered by using these instruments.

For a more detailed explanation of the use of modern instruments reference should be made to:—

Electro-Magnetic Distance Measurement by C. D. Burnside M.B.E., B.Sc., A.R.I.C.S. — Aspects of Modern Land Surveying by J. R. Smith A.R.I.C.S:

and, in addition, the instrument manufacturers' technical literature.

A. Electromagnetic Distance Measuring (E.D.M.) Instruments

Perhaps the best known instrument of this class used in recent years is the DISTOMAT manufactured by Wild of Switzerland. Setting this instrument over a point and measuring to a target up to 2·5 km distant, it is capable of an accuracy of ± 10 mm. The versatility of such an instrument is clearly apparent.

However, the use of them to date has been limited to major projects, but there are many aspects of small and complex projects where their use could be justified. These would include situations where accurate taping is impractical or time-consuming; for example, a large number of manhole positions on a hilly site or the principal points for centre line geometry. The difficulty in using conventional instruments and tapes in heavily built-up urban areas, particularly where traffic interrupts measurement, would be further situations where E.D.M. instruments could be used to advantage.

There are a number of instruments comparable to the DISTOMAT. These include the Tellurometer, AGA 700, Zeiss Reg Elta.

B. Direct Reading Optical (D.R.O.) Instruments

These instruments are a great improvement over the conventional tacheometric methods of surveying. They provide direct distances both horizontal and vertical, without the use of any conversion tables. They are invaluable for the rapid determination of spot levels and surveys for earthwork quantities. They are not quite as accurate over long sights as the E.D.M. instrument.

Examples of these D.R.O. instruments are the "Wild R.D.H." and Kern DK−RT.

A further refinement is the gyro theodolite. These are used in complex tunnelling and mining where they can determine direct geographic bearing from true North to an accuracy of ± 30 secs.

C. Lasers

The laser beam provides exact horizontal and vertical alignment. The instrument is used more for the control of the accuracy of on-going construction work rather than pre-determining setting out points. A good example is the laser control of slipform motorway paving machines and the alignment of pipelines. Lasers are also being used to control the verticality of structures in the place of more conventional methods of plumb bob and optical plumbing instruments.

PART 2
Pipelines and Sewers

I. **INITIAL CHECKS**

Before commencing any setting out work, the site engineer must walk along the planned line of sewer and systematically note any features likely to cause problems in setting out, or actual conduct of the work.

The site engineer must refer to all the available information and check for the points listed below. Time spent on this and "walking the line" is never wasted and may well help to prevent standing time and lack of progress at a later date.

A. Distance Between Adjacent Sewers

Check with site management that the distance between adjacent sewers shown on drawings will permit the use of the size and type of excavator that it is proposed to employ. Size of pipes and ground conditions will affect this.

B. Interference By Drain Connections

Inspect the drawings for the location of connections, and determine whether these will interfere with any adjacent sewer runs, particularly as regards level.

C. Crossing Existing Services and Sewers

If, from the drawings, the new sewer line crosses existing mains and sewers, find out the reduced level of these services. Normally, the most reliable way is to dig trial holes, but instructions from the engineer and site management must be obtained first.

It is dangerous to assume that the level of a point on the line of an existing sewer can be calculated by working out the gradient between two manholes. Do no forget to take into account pipe thickness, as this may be important in "tight" spots.

D. Manhole Positions

When "walking the line", agree and confirm with the engineer the actual position of the manholes on the ground. Normally, sewer drawings do not show exact positions of manholes or take account of small local features and obstructions.

E. Manhole Clearance

Make sure that when two sewers are laid side by side there is adequate clearance between the outside of the manhole wall and the adjacent sewer.

F. Services

Consult the drawings showing public utilities' mains and cables to make sure that the proposed sewers do not interfere with them, particularly the manholes. If necessary, dig trial holes.

G. Discharge Levels

If a new sewer is to discharge into an existing manhole or other outfall of fixed level, check physically on site that the level shown on the drawings is accurate. This is very important, as errors in this respect may mean re-design of part or whole of the system.

H. Gully Connections

Check that the surface water sewer invert is such that connections from road gullies can be made. The distance below the top of the cast iron gully grate to the invert of the outlet pipe from the gully pot will normally be about 675mm (allowing for two courses of brickwork).

If, as a result of these initial checks, it appears that discrepancies and/or errors are present, they must be taken up and cleared with the engineer before proceeding with the setting out.

II. SETTING OUT PROCEDURE — LINE AND LEVEL

Before setting out for pipe lines or sewers, the level of any temporary bench marks it is proposed to use must first be verified.

All relevant information must be recorded on a sewer information sheet or general purpose information sheet and copies made available to all concerned. Do not rely on verbal instructions only.

A. Centre Line

The centre line of the trench is first pegged out by the site engineer using white coloured pegs, as described in Part I. The distance between pegs should be limited to 30m with a peg at the centre of each manhole, referenced to other pegs or nearby features.

B. Off-set Pegs

When the centre line has been pegged out, off set pegs are positioned between 3m and 6m to the right or left of the centre line looking up the flow. The actual distance from the centre line can be varied to suit the amount of excavated material and the requirements of the plant being used for excavation and transport. The off set distance must be kept standard between manholes. Offset pegs should be coloured yellow and driven to within 50mm of ground level.

C. Profiles and Travellers

The standard sewer profile described in Part I must be set up over the off set pegs. The position of the end of the profile board must be about 3m from the centre line depending of course upon the distance chosen for the off set pegs.

The top edge of the profile board must be fixed in relation to pipe invert levels and values selected to allow the use of travellers with a length forming a multiple of 250mm e.g. 2·5m. Details of the construction of standard travellers for sewers are described in Part I.

The depth of trench is first calculated, the length of traveller chosen and the distance from peg to top of the profile board calculated. The profile board is then positioned by measuring up from the peg as shown in Fig. 12.

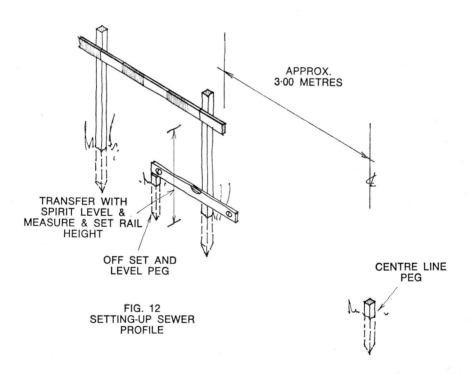

FIG. 12
SETTING-UP SEWER PROFILE

Only in exceptional circumstances should levels be set which require a traveller length which is not a multiple of 250mm. If possible, avoid setting levels which require a traveller of less than 2m.

It is good engineering practice to provide at least three profiles for each length of trench, one at each end and one in the middle of the run. The object of the centre profile is to allow quick checks to be made upon the continuing accuracy of the other two .The level for the centre profile must always be worked out from the level and gradient shown on the drawings. Never set the centre profile level by boning through from the end profiles.

Finally, it is emphasised that only the consistent use of a sewer information sheet will ensure that foremen and gangers have a clear indication of what is required with each length of sewer.

III. MANHOLES

Each manhole must be separately detailed by the site engineer on a general purpose site information sheet. Special attention must be given to any divergencies from the typical manhole construction used on the job. Very often separate gangs of operatives from those used on the pipe work are used to construct manholes. It is as well therefore to give details of manholes separately to the gangers involved, including the referencing.

The concrete rings should be given a code letter depending upon their depth and type on the sketched detail. The site engineer should also paint this code letter on the rings when they have been off loaded. This helps to ensure that rings are placed in the right order and the manhole is the correct depth.

When drawing up the manhole details, the site engineer must work from a check list similar in form to that on the sewer information sheet. Some of the items on the list would be, for example, details of:

1. Base
2. Benching — including channels
3. Taper
4. Concrete surround
5. Cover slab level

The temptation to complete manhole construction well behind the construction of the pipe works should be resisted. Experience has shown that on most sewer jobs the completion of manholes are close to the completion of the pipe work as practicable has paid off in terms of cost savings and quality of workmanship.

IV. JUNCTIONS

A. Marking Position Before Laying

The sewer information sheet must show the plan position of junctions on the pipeline. The dimension is given to the ganger in numbers of pipe lengths taken from the inside face of the manhole wall or ring. Remember, that some specifications require a cut pipe as the first pipe out of the manhole. In these cases, express the length as a fraction of a pipe plus whole pipes, the *last whole pipe* being the pipe with the junction.

The position for saddle junctions where fitted later must be measured by tape.

The angle of junctions must be estimated and described as Flat, Medium or Steep as shown in Fig. 13.

Make sure that general foremen and gangers understand what is meant by these terms by providing them with a reference drawing as Fig. 13.

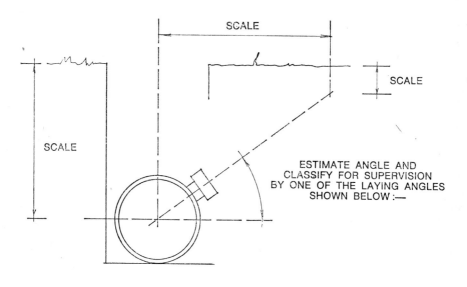

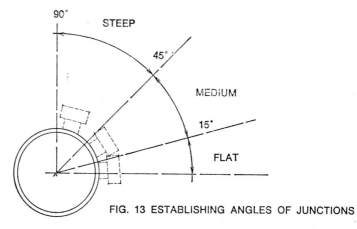

FIG. 13 ESTABLISHING ANGLES OF JUNCTIONS

B. Marking the Position Before Backfilling

The site engineer must see that whoever is responsible for backfilling, normally the general foreman or ganger, fastens a stout piece of wire to each junction attached to a 1·5m length of 225mm x 75mm timber. The timber can then be placed upright, if depth of trench permits, against the side of the trench before backfilling. The "as fixed" position of the junction should be measured upstream from the lower manhole by the site engineer and recorded on his copy of the sewer information sheet.

Where the trench is in existing hard road construction, and temporary reinstatement of the trench has to be carried out, the position and direction of the junction can be cut into the road surface at the edge of the trench. These marks will be removed during the permanent reinstatement. A timber marker should still be placed in the backfill as described above.

V. BACK DROPS

When dealing with back drops in a stretch of sewer, errors often arise due to the abrupt changes in depth. In order to avoid this, the following procedure should be adopted:
1. Record position on the sewer information sheet
2. Set up the profile posts for the section containing the back drop, but do not fix the cross piece until the preceding section is excavated.
3. When the back drop section is ready for excavation, withdraw the traveller, re-set the profile and issue the new traveller. Details must be shown on the information sheet.
4. Show details of the construction of the back drop and manhole to the general foreman and ganger on a general purpose information sheet.

VI. DUAL SEWERS IN DUAL TRENCHES

A typical cross section of a dual sewer and trench is shown in Fig. 14.

The following procedures must be adopted:
1. Inform the general foreman and ganger of requirements and issue a sewer information sheet.
2. Erect two sets of profiles as shown and paint the sight rails red and white for foul sewers and yellow and white for storm water.
3. Construct the traveller with one cross piece which will be used in conjunction with the upper or lower sight rail as appropriate.
4. Check that the invert levels of the sewers will allow over and under cross connections to the adjacent pipes.

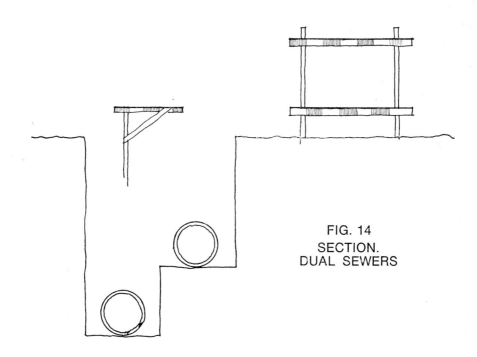

FIG. 14
SECTION.
DUAL SEWERS

VII. CONCRETE BED

The modification of the traveller for the concrete bed thickness is described in Part I, Fig. 6. Details of bed thickness and length must be shown on the information sheet. Make sure that the ganger understands the specification for the bed. The site engineer must ensure that the dimensions of the bed are as far as possible those specified. It is too easy for the full width of trench to be filled with concrete if control is not exercised.

VIII. DEEP SEWER TRENCHES

The procedure for setting out deep sewer trenches will depend entirely on the methods adopted and the plant to be used for excavation. In general, the level should be controlled in stages, the limiting factor being the length of the traveller. For easy handling, it should not be more than 5·00m long.

If the sides of the excavation are battered, then the profiles must be erected as shown in Fig. 15.

When the excavation is within 250 mm of the bottom of each stage, profiles must be erected at the bottom of the excavation and work continued using the new level references. Profiles can be fixed on to the timbering, but the site engineer must check that this is not liable to movement. Where sheeting is being driven progressively, profiles must be dismantled whilst driving is in progress and re-erected after when there is no danger of the timber moving.

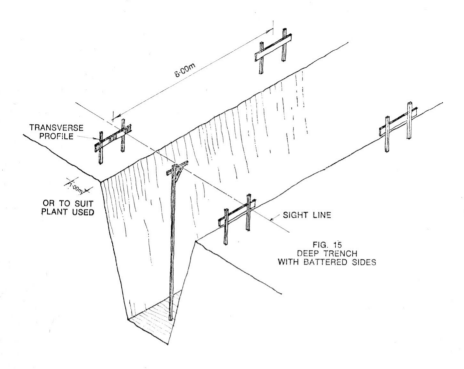

FIG. 15
DEEP TRENCH
WITH BATTERED SIDES

PART 3
Roadworks

I. INTRODUCTION

This part of the manual deals with the setting out of roadworks, and is divided into four sections. The first section deals with the general procedures which should be adopted for all types of roadworks. The next section deals with the main setting out features encountered on housing and industrial estate roads. The third section deals with the special requirements for main highways.

The last section contains worked examples of setting out using co-ordinates, and calculations for vertical, horizontal and spiral transition curves. Employing authorities are nowadays making increased use of computer programs to design highways schemes. Information, including co-ordinates, is supplied to the contractor in the form of computer printouts and an example of such a printout is given.

This part of the manual deals with normal setting out procedures and instruments which may be used in typical construction projects, and should be within the capability of site engineers. Information on more specialised instrumentation is contained in Part I, Section VI of the manual.

II. GENERAL PROCEDURES

The following procedures are applicable to all types of roadworks. They should also be read in conjunction with the general procedures regarding setting out contained in Part I of the manual.

A. Proving the Survey

It is always necessary to 'prove' a survey before starting setting out work. The object is to check that the client's layout will fit the site, particularly in relation to road lengths which may be inaccurate on the drawings. The site engineer must verify that the main setting out stations and their co-ordinates have been checked on site in advance.

Survey proving should be carried out as follows:
1. Establish intersection points on curves.
2. Measure the intersection angles.
3. Determine lengths of tangent straights and chainages. Check these against the drawings and/or client's calculations.

4. Check the position and co-ordinates of intersection points and tangent points by triangulation to the main setting out stations.

   Check also chainages at hedge and ditch crossings.

5. Run a series of check levels along the centre line at 30m intervals and check the results against the drawings.

Natural features can be of considerable assistance in checking a set out. The site engineer should always bear this in mind and use them wherever possible to verify his work. Natural features are, however, frequently only plotted approximately on surveys, especially the less permanent ones such as trees and hedges. Site engineers should therefore always check the accuracy of an important reference point, i.e., a tangent point on a curve or the terminal point on a length of straight road, by measuring to at least two or three clearly established natural features. Use the methods described in Section V.A. — Setting Out using Co-ordinates.

If no survey is available, it will be necessary to establish the setting out stations. As this is a costly procedure, it is always worth enquiring from the client's representative whether or not an original survey has been made of the site. Sometimes this sort of information is contained in the title documents of the owner of the freehold.

If it is necessary to establish setting out stations, the following procedure should be followed:

1. Refer to the largest scale and latest Ordnance survey available for the area.

2. Select recognisable permanent features like church towers, particular buildings, groups of rocks etc. Check these are still visible on the ground.

3. By triangulation or traverse from these features set up stations on the site. Use methods outlined in Section V.A.

4. Use the TBM metal stake and concrete surround described in Part I to fix the stations.

5. Try to use these stations as combined TBM's and setting out stations. The setting up of these stations is costly in time and money — it is important that they should be protected as shown in Part I.

6. Remember to position stations, so that future construction activities will not hinder their intervisibility. For example, spoil heaps, sites offices, and the early building operations could obscure visibility. Check with the site management.

When the reference stations have been set up, the site layout should be "proved" as described above.

The engineer or his representative must be informed of the position and co-ordinates of setting out stations. The site engineer should arrange for a member of the engineer's staff to accompany him when setting up reference stations and to check the measurements.

B. Dealing with Discrepancies

If discrepancies are discovered when proving the survey, the site engineer must make quite certain that the drawings are in error before bringing them to the attention of the engineer.

The site engineer must always, in conjunction with site management, suggest solutions which can be submitted to the engineer. The object should be to minimise extra costs and delays to the contract.

Some common discrepancies, and the action which should be taken, are as follows:

1. The road lengths on the drawings are not compatible with the proving on the ground.

   After the engineer's representative has checked the proving set out, amendments will have to be made to the drawings, particularly the distances between cross-sections which in turn will alter the levels.

   A great deal of confusion can arise here, unless a new longitudinal section of the correct length is drawn together with fresh cross-sections showing the new road gradients. This will normally be the responsibility of the engineer, but the site engineer should give all the assistance he can in order to avoid delays in starting the work.

   The positions of tangent points may have to be adjusted to suit corrected measurements and fresh curve data will need to be obtained from the engineer.

2. The intersection angles, as measured on site, are not the same as shown on the drawings. Refer this to the engineer's responsibility.

   Four courses are open:

   a. Maintain the curve radius as shown on drawings and alter the tangent lengths to suit the intersection angle.

   OR

      b. Alter the radius and leave tangent lengths as shown.

OR

      c. Alter the direction of the tangent lengths and then by trial and error try to obtain the intersection angle shown on the drawings. Then use the tangent lengths and radius data given originally.

OR

      d. Adjust the straight lengths between adjacent tangent points.

3. "Spot levels" differ greatly from levels shown on drawings. In this case, the site engineer must discuss the implications with site management and then the engineer's representative. It may be necessary to re-design some of the road gradients to avoid excessive cut and/or fill.

C. National Grid and its Application

A great many schemes prepared by the Department of the Environment Road Construction Units and Development Corporations are based on the National Grid.

Briefly, the National Grid is the projection for this country on which all Ordnance Survey plans are constructed. It is based on the primary triangulation of Great Britain, which in turn is sub-divided into 2nd, 3rd and 4th order triangulation.

The triangulation stations are usually pillars specially erected, marks on church towers, water tanks, buried concrete blocks and church spires.

The description and National Grid co-ordinates of such points may be obtained from the Director General, Ordnance Survey, Southampton, on payment of the appropriate fees.

The distances calculated from any two Ordnance Survey stations are projection distances and differ from physical distances (i.e. actual ground distances measured). To convert projection distances to ground distances, it is necessary to use a Local Scale Factor.

$$\text{Thus the ground distance} = \frac{\text{Projection Distance}}{\text{Local Scale Factor}}$$

The table below gives the value of the Local Scale Factor for any part of the country.

National Grid Easting (Km)		Scale Factor F.
400	400	0.99960
390	410	60
380	420	61
370	430	61
360	440	62
350	450	63
340	460	65
330	470	66
320	480	68
310	490	70
300	500	72
290	510	75
280	520	78
270	530	81
260	540	84
250	550	88
240	560	92
230	570	0.99996
220	580	1.00000
210	590	04
200	600	09
190	610	14
180	620	20
170	630	25
160	640	31
150	650	37
140	660	43
130	670	1.00050

Example:
O.S. Station   National Grid Co-ordinates
O.S. Station   E. 482841.570   N. 263542.210
                E. 478835.970   N. 261889.570

Difference   4005.600   1652.640

Projected distance = $\sqrt{4005.600^2 + 1652.640^2} = 4333.134$
Physical distance = $\dfrac{4333.134}{0.99968} = 4334.521$

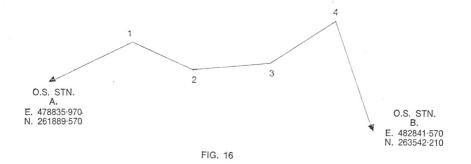

FIG. 16

(i) Application of National Grid to traverse

To calculate a traverse A—B (Fig. 16) it will be necessary to apply Local Scale Factor (L.S.F.) to all measured distances before traverse is completed. (National Grid projections do not necessitate any correction to measured angles or bearings.)

Line	Hor. Distance of Measured Line (m)	L.S.F.	Dist. Projected (m.p.)
A—1	117.352	0.99968	117.314
1—2	203.891		203.826
4—B	256.480		256.398

(ii) Application of National Grid to setting out

National Grid Co-ordinates of A, B, C, E (Fig. 17) given on computer print out.

To set out the principal points using the National Grid Co-ordinates given on a computer and National Grid Co-ordinates of Control Traverse:

National Grid Co-ordinates of 1   E. 478650.899   N. 264968.753
National Grid Co-ordinates of A   E. 468522.603   N. 264901.320

$$ 148.296 67.433$$

Projected Distance 1—A $= \sqrt{148.296^2 + 67.433^2} = 162.908$ m.p.
Setting out Distance 1—A $ 162.908 = 162.960$ m.

$$ 0.99968$$

BRG. 1—A $= TAN^{-1} \dfrac{148.296}{67.433} = 245° 32' 52''$

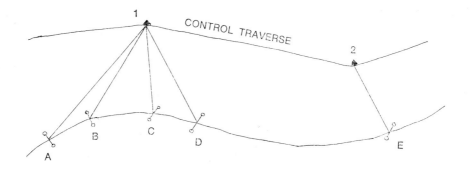

FIG. 17

Setting out Main Centre Line from computer print out

Whilst no correction is necessary to the bearings and angles, all distances on the print out are projection distances and have to be divided by the appropriate L.S.F. to obtain the equivalent ground distances for setting out.

For further detailed information regarding the National Grid, reference is made to the Ordnance Survey publication:

"CONSTANTS, FORMULAE AND METHODS USED IN TRANSVERSE MERCATOR PROJECTION", obtainable from H.M. Stationery Office.

D. Accuracy

A high degree of accuracy should be maintained when setting out roadworks. Road centre lines should be set out to an accuracy of better than ± 25 mm from true plan position in every 100 metres. This will depend upon the configuration and lengths of roads. One would certainly not expect a cumulative error of 75 mm. The worked example in Section V was set out from one point through to the opposite extreme point and the traverse was closed to within 50 mm—this is the sort of accuracy to aim for.

Circular and transition curves should be set out using a theodolite reading to 20 seconds of arc or less.

Road lengths are critical, particularly on the estate roads, and, to ensure that frontages are correct, they should be determined by measuring along the centre line after a length of road has been set out and checking against the lengths shown on the drawing.

E. Keeping site staff informed

The routine preparation of information as described in Part I is essential so that the possibility of misinterpretation by site management or supervision is avoided. Information should be recorded by the site engineer on General Purpose Information Sheets and distributed to all concerned. In addition, verbal explanations should be given to general foremen and gangers. The frequent use of clear sketches to illustrate the point cannot be over emphasised.

Make sure that all the information needed to determine the main setting out points shown on site plans and a list and sketch of key stations and TBM's is displayed in all site offices.

III. HOUSING AND INDUSTRIAL ESTATE ROADS

A. Initial Checks

Very often plans and sections for this type of work do not contain all the essential information that the site engineer will need for setting out the work on the ground. It is important therefore that the drawings are carefully examined and that errors and/or omissions are recognised and cleared with the engineer's representative before any work is started on site.

The following points must always be examined on estate roads drawings:

1. **Curved Radii**
   Curves may have been drawn freehand or with French curves. If necessary, design a suitable curve using the pro-forma shown in Section V.C.

2. **Reference Points**
   If none is shown, and there is no grid reference, decide suitable permanent reference points on site and establish stations as described in Section V.A.

3. **Levels on Sections**
   Particularly those concerning gradients, vertical curves and junctions.

4. **Positions of Road Gullies**
   These may not be shown on the plan in the lowest part of the road, particularly on vertical curves. Where gullies occur at road junctions, make sure that they will drain both roads, allowing for camber and/or cross fall. It is often necessary to draw the junction to a large scale, plot on a grid of levels and verify that the water will run off all parts of the road to a correctly positioned gully.

The site engineer must bring to the attention of the engineer's representative any errors and omissions, so that alternative instructions can be issued in good time.

B. Standard Setting Out Procedure—Line

1. **Centre Lines**

    The site engineer must position all centre line pegs. They must be driven to about 50 mm above ground level and painted white as recommended in Part I. The pegs must have a wire nail driven in the top for sighting and for a steel tape to be hooked over while offsets are being fixed.

2. **Off-set Pegs**

    It is good practice to set off-set pegs a standard distance back from the kerb face—say 1 metre. When footpath and/or verge excavation is to be done at the same time as the road excavation, set the pegs 1 metre from the back of the path line.

    There will be exceptions to the standard depending on site conditions and the type of earth-moving plant to be used. Before setting the pegs, find out the minimum width in which the earth-moving plant will operate successfully. If the footpath is narrow, it may pay to over excavate the width in order to construct the path. Consult site management over these points and set the pegs 1 metre behind the excavation line. Make sure that the information is recorded on a site information sheet.

    Right angles for off-setting from the centre line can be set out quickly and with sufficient accuracy with an optical square.

    Off-set pegs must not be concreted in position, but a nail should be driven in the top to denote exact off-set measurement from centre pegs. Use the recommended colour-coding for all off-set pegs.

3. **Reference Points**

    Many important reference points such as tangents and intersection points will sooner or later in the construction process be displaced either by accident or design. It is essential therefore that these points can be re-established quickly and accurately at any time.

    In many congested sites it is just not possible to locate offset or reference pegs at convenient distances from important setting out points. In these circumstances position a series of reference pegs equi-distant from the main setting out point. After setting out intersection and similar points, take bearings and distances on to at least three of the reference pegs for each point. Record this information carefully, and file it in the site office for later use as necessary. The reference pegs must be surrounded with concrete as for the TBM's and protected.

If there are no obstacles, and site operations are not going to extend too far, then a simple system using three pegs and a standard dimension of, say, 15 metres to the point can be used, Fig. 18.

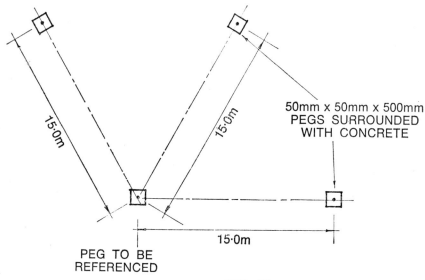

FIG. 18
ESTABLISHING REFERENCE
POINTS WHERE SITE OPERATIONS
ARE LIKELY TO DISRUPT MAIN PEGS

The site engineer must always be alive to the day-to-day construction needs and anticipate the likely disruption of setting out points. He should always be in a position to re-establish quickly any point. The adoption of the recommended system of colour coding of pegs will assist him to identify the function of each peg on site.

C. Standard Setting Out Procedure — Level

1. Temporary Bench Marks (TBM's)

All TBM's should be set up as described in Part I. If no permanent features are available to serve as a TBM, combine the function of site setting out reference stations with those of a TBM.

TBM's should be placed in positions which will not be impeded by the construction of the roads nor by any later building operations.

2. **Cross Sections**

By carefully studying the drawings it is usually possible to position centre line and offset pegs to coincide with the level points shown at cross and longitudinal sections. Points where levels are required on the ground, but not shown on drawings, must be calculated by the site engineer using what information there is on the drawings and from nearby levels and gradients. All such calculations must be shown on a site information sheet, and filed for reference purposes.

It is unusual for all level information to be available on drawings for estate roads, and some calculation will usually be necessary. To ensure accuracy in excavations and succeeding operations, levels should be established at:

a. Changes in gradient.
b. Changes from a straight cross fall road to a cambered road.
c. 10 metres intervals on vertical curves.
d. The lowest point in a "sag" in a vertical curve.
e. Distances not exceeding 20 metres on straight lengths of roads.
f. Distances not exceeding 15 metres on horizontal curves.
g. Distances not exceeding 20 metres on horizontal curves over 150 metres radius.

3. **Profiles**

A suggested form of profile for roadworks use is shown in Fig. 19. The cross piece must always be set parallel to the centre line of the road. Boning between profiles with a traveller can then be done more accurately. This is particularly important when there are crossfalls and cambers to form.

The site engineer must make himself personally responsible for setting up all the profiles. Profiles should be set up in pairs, one each side of the centre line. They should be erected to coincide with centre line offset pegs and positioned just behind the offset pegs. The site engineer should enter calculations for level, traveller length, and position on a site information sheet and give copies to the general foreman and ganger.

The information sheet should show:

1. Reduced level of the top offset peg being used with each profile.
2. The chainage (metric).
3. The chosen length of traveller (see note in Part I).
4. The reduced level of the top of the profile board, the length of traveller having been decided first.

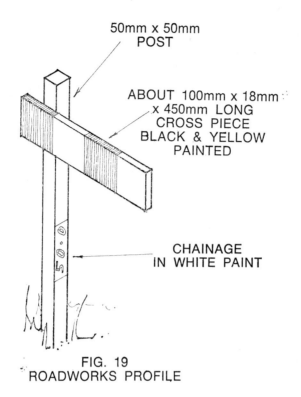

FIG. 19
ROADWORKS PROFILE

5. The distance in millimetres from top of peg to top of profile board.

The site engineer must mark in black waterproof pen on the profile post the reduced level of the profile board.

It is also of advantage, in addition to marking the level of a profile, to mark the chainage. This is best done on the profile post using white paint. Set the level of the rail by levelling across from the offset peg with a spirit level and using a levelling staff to show the distance to be added to the peg reduced level.

Roadworks profiles are particularly vulnerable to damage and displacement by the construction process. Daily checks on profiles are necessary or more frequently when heavy plant is being used in the vicinity. Do this by "sighting" over a line of profiles. Site engineers must be prepared for fairly frequent renewal of profiles.

Whenever possible, profiles must be set 1m above road channel levels. This is after making appropriate adjustments for crossfalls and distance of the profile from the channel.

4. **Travellers**

The recommended form of roadworks traveller is the self-supporting adjustable type, described in Part I, Fig 4.

Before setting the length of traveller, the site engineer must find out from site management what allowance will be made for compaction of the formation after rolling. This allowance will vary from site to site depending upon specification and tolerances, if any. The height of traveller will then be set the required allowance (in millimetres) short of the finished formation. It is all too easy to confuse general foremen and gangers over traveller lengths. Therefore, the length of traveller is set after allowing for compaction, and no further information need be given, Fig. 20.

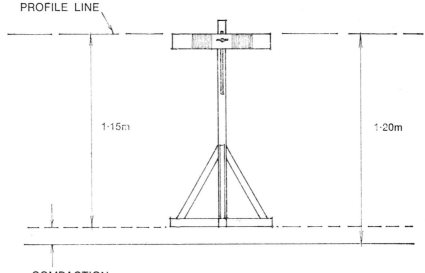

FIG. 20 COMPACTION ALLOWANCE ON ROADWORKS TRAVELLERS

D. **Footways**

When footways are 1.5 or 2m wide, excavating to the correct crossfall with earth-moving equipment will be difficult, particularly if scrapers are used. It is therefore necessary to excavate to a "bone" between profiles both longitudinally and transversely. The fall across the path can then be trimmed by hand afterwards. Excavation is carried out first to footway formation level only across both roadway and footway, Fig. 21.

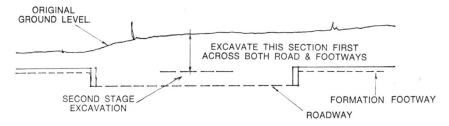

FIG. 21 EXCAVATION OF ROADS & FOOTWAYS

Profiles are then transferred and set 1m behind the line of the kerb face, taking measurements from the centre line offset pegs. The level of the top edge of the profile cross piece should be 1m above the channel level. Road excavation can then proceed to road formation level.

The site engineer must remember that different length travellers will be required for footway and road excavation.

E. Cambers

When a camber is required in the formation of the road excavation, the site engineer should have the traveller modified. The cross piece should be made up from timber equal in height to the required camber. The upper edge of the traveller will correspond to the channel level and the lower edge to the crown of the road formation, Fig. 22.

An alternative method is to use two cross pieces on each profile set so that by using a traveller and sighting from the upper to the lower cross piece, and vice versa, across the road the true camber is obtained. Always mark on the profiles which sight line is which to avoid any confusion.

The site engineer must make sure the general foreman and gangers are properly instructed on the use of these travellers. Use a sketch similar to Fig. 20 on the information sheet to demonstrate the use. This becomes increasingly important where cambers vary on one job or where they change to crossfalls. The site engineer will need to check the use of the travellers frequently and to inform the site supervisor in good time when changes in levels will be needed.

Scratch boards and camber boards, in addition to travellers, can greatly assist in the accurate forming of the road fill or sub-base when the kerb or channel has been laid.

On roads over 5.5m wide these boards are difficult to handle and, as a result, will tend not to be used. In such cases, level pegs should be put in not more than 15m apart along the road centre lines and intermediate pegs at 3m intervals boned in between.

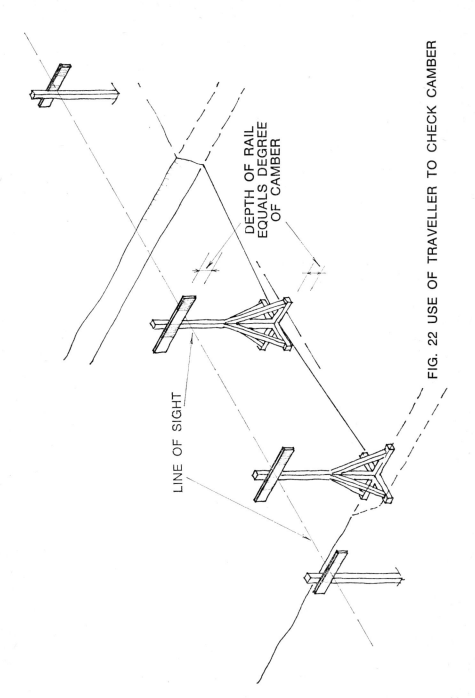

FIG. 22 USE OF TRAVELLER TO CHECK CAMBER

The camber board or scratch board can then be made up for half the road width, Fig. 23.

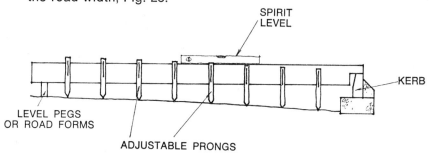

FIG. 23 CAMBER & SCRATCH BOARD

The control of fill to footways can be done most easily when the path edge has been laid. Use boards as shown in Fig. 24 below, so that the whole of the footway area can be measured and controlled.

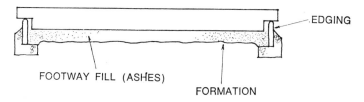

FIG. 24 FOOTWAY CAMBER BOARD

F. Kerb Levels

The control of levels for kerb laying is best done by putting level pegs 0.50m behind the face of the kerb. The peg should be driven in until the top is at kerb reduced level. These pegs should not be more than 15m apart on straight runs or on horizontal curves and not more than 8m apart on vertical curves.

G. Channel Levels With Valleys and Summits

Where the road gradient is flat, usually flatter than 1 in 250, channels will have to be laid to valleys and summits. More kerb face will show at a low point where the gully will be sited, and less will appear at a summit. To achieve this, the site engineer should examine the drawings and specification early on in the contract, making sure that the gully positions reconcile with the proposed kerb face and gradient details.

When the kerbs have been laid, the site engineer should mark out in yellow waterproof chalk the level and position of each "valley" and summit using a level and staff. The kerblayer can then "snap" his chalked kerb line between these points to give accurate levels for the short lengths of channel to be laid.

A sketch of a typical length of kerb with maximum and minimum kerb face detail and distance between should be prepared on an information sheet and a copy handed each to the general foreman, ganger and kerblayer concerned. A copy should be kept for site office records.

IV. MAIN HIGHWAYS

A. General

The principles of setting out the main centre lines and important reference points in main highways are the same as those outlined in the worked example in Section V.A. The main setting out stations are often provided by the employing authority and the co-ordinates of the stations should refer to the Ordnance Survey grid. The task of checking these stations is specialised work and would not normally be the responsibility of the site engineer. In principle, this work is no different from that outlined in Section V, but instrumentation is more specialised (Part I, Section VI).

The main stations are usually substantially constructed reference pegs mounted in a large block of concrete. They are normally some distance from the roadworks operations, and the site engineer will need to construct site stations closer to the site. These site stations will be referenced back to the main stations by triangulation.

Before setting out is commenced, the survey should be proved as described in Section II and any discrepancies brought to the notice of the engineer immediately.

B. Establishing the Centre Line

When the proving survey has been completed, the centre line should be set out in detail. This set out will follow the general procedure outlined in Section V. Generally, the order in which the centre line will be set out is as follows:

1. Establish intersection points.
2. Measure and record intersection angles.
3. Check lengths back to reference station co-ordinates.
4. Establish tangent points of transition and circular curves by chainage from intersection points.
5. Set out by theodolite deflection angles for transition and circular curves. "Steel bands" must be used to measure chord lengths. Nails should be driven into the standard

colour-coded setting out pegs to mark exact points and facilitate tape handling.

On contracts with transition curves it will be necessary to set out each kerb line separately and then off set in addition to the centre line.

Once the centre line has been established, the following can be set out from it before work starts :

1. Fence lines.
2. Drains and ditches.
3. Site stripping area.

Main chainage off set pegs should be placed 1m behind the limits of site stripping and embankment filling.

The offset distances from the centre line should be standardised as much as possible and all information recorded on a site information sheet.

Site engineers must remember that construction work invariably disturbs lines. When embankments or cuttings have been formed, the centre line must always be re-established.

Centre lines must always be checked and transferred before kerbs and channels are laid.

C. Establishing Existing Ground Levels

Before site stripping starts, the site engineer should check and record existing ground levels at each cross-section. If there is a marked change in level which does not appear on the existing cross-sections levels, additional cross-sections must be established, recorded and agreed.

D. Widening Existing Highways

When work consists of widening an existing carriageway, the surface of this is used as the setting out line.

Road pins should be driven in and used as the permanent line from which to set out off sets.

Pins should be numbered in white paint on the road surface and a list of them prepared on the information sheet. The list should show :

 a. Pin number.
 b. Chainage.
 c. Off set distance required to a point 1m beyond face of new kerb.

Road pins should not be used for level. It is better to provide separate standard pegs for this purpose.

E. Cuttings

1. Setting Out

The following procedures should be adopted :

1. Take a series of ground levels along each cross-section and any intermediate points where profiles will be required.
2. Plot the reduced levels on the cross-sections shown on the drawings.
3. Establish the approximate width of the cutting by scaling off the drawing from the highway centre lines.

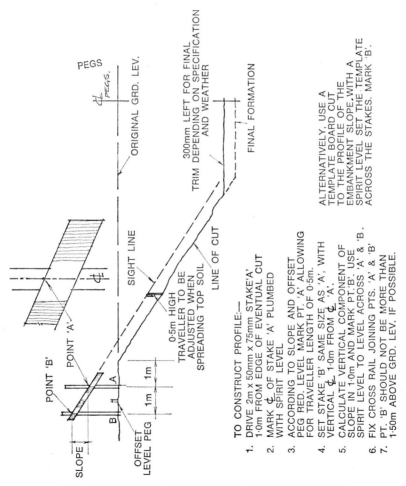

FIG. 25 CONTROL OF CUTTINGS AND EXCAVATION

45

4. Calculate the true width of cutting and the position and level for the profile as shown in Fig. 25, keeping the stake 'A' 1m from the edge of the cutting. For the most usual batters found in highway work a traveller 0.5m long will be sufficient. Remember that this traveller will need to be modified when topsoil is to be spread, or when fresh profiles are erected. Profiles should not be more than 50m apart longitudinally and correspond with the centre line chainages.

2. **Controlling the Work**

    The site engineer must maintain close contact with the excavation work as it proceeds, checking the centre line frequently. Use ranging rods to show the centre lines, as pegs will be difficult to see by the plant operators. It is usual to leave about 300mm of sub-soil unexcavated above the required formation level in cut and overfill by 150mm in fill areas. This will vary to suit the type of sub-soil and weather conditions prevailing, but it is vital that the formation is sound when the remaining second stage is finally removed. When the excavation is down to within 300mm of formation, standard roadworks profiles and travellers should be used to control the trimming and grading of the formation to crossfalls and cambers.

F. Embankments

1. **Setting Out**

    The procedure for embankments is similar to that required for cuttings. The profiles are constructed in the same way as shown in Fig. 26, keeping point 'A' about 0.5m above ground level. This is to facilitate sighting up the embankment. A traveller of 1.5m is sufficient for most embankments found in highway work. Where embankments exceed 5m in height, construct further sets of profiles on the slope at intervals of 5m.

2. **Controlling the Work**

    It is essential for the site engineer to maintain close contact with the works and to see that uncontrolled overfilling does not take place. The centre line should be checked regularly. The layers of fill and consolidation are controlled by placing temporary uncoded level pegs at the outside extent of of each layer—suitable allowances being made for consolidation.

    It is customary to over-fill by about 150mm. Profiles and travellers are then used before the removal of the second stage and trimming to formation.

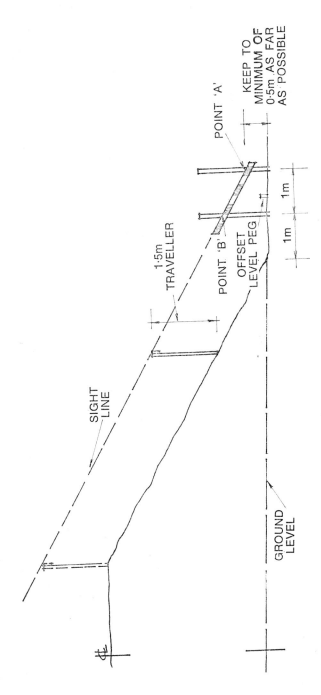

FIG. 26 CONTROL OF EMBANKMENTS & FILL

G. Land Drains

Herringbone land drains under the formation of the road are often required. The requirement in positioning these drains is usually not great, and most of the setting out can be left to the ganger and general foreman.

The site engineer should provide the ganger with a sketch of the drain layout with dimensions of each network. Levels should be established by the site engineer by means of pegs. The ganger can take measurements and transfer pegs to the bottom of the trenches by means of a 4 metre levelling board and spirit level.

The lines of the pipes can then be set using steel pins and bricklayer's lines. Periodic checks on dimensions, as laid down, should be carried out by the site engineer.

H. Spoil Heaps

The correct positioning of spoil heaps in relation to the excavation and subsequent construction of highways is of vital importance. The whole field of optimum haulage routes for scrapers and dump-trucks to spoil heaps has been the subject of specialist study in recent years. There are computer programs available, specially written to solve the complicated mathematical relationships which exist when optimising length of haul against size and type of machine and size and shape of the eventual spoil heap. Site management will have decided on the economic displacement of spoil heaps and haulage routes before excavation work starts.

The site engineer should set out the plan position of the spoil heap on the ground using 2m stakes painted white. The plan and profile of each heap should be sketched on a standard information sheet and given to the general foreman and ganger concerned.

Haulage routes will vary a great deal depending upon soil conditions and the frequency at which they are used and the weight of traffic using them. For the more elaborate heavily used routes, substantial construction using various grades of fill materials will be needed. For these, the site engineer will be required to set lines and level pegs and control the construction in a similar way to the more permanent roadworks.

All spoil heaps and haulage routes should be clearly marked by direction boards and arrows large enough for operatives of heavy plant to read them from their cabs.

V. SETTING OUT CALCULATIONS

This section contains worked examples of the usual calculations required for setting out roadworks, together with an explanation of the procedures to be followed.

A. Setting Out Using Co-ordinates

Figure 27 shows part of the layout plan of roads to a housing estate. The main setting out stations and their co-ordinates have been provided by the engineer. These stations are Nos. 8, 13, 10 and 5. All the co-ordinates refer to a local 100m grid starting on this part of the site at 9,200m East and 3,900m North.

The site engineer is required to set out the centre lines of two roads, numbered 34 and 31.

The first procedure for the site engineer is to walk over the site and visually inspect the station pegs, checking for any obvious signs of damage or displacement. If any pegs have been disturbed, this must be brought to the notice of the engineer's representative.

Note the position and station number of any suspicious station pegs. It is essential for all station pegs to be properly protected from site operations; this is the contractor's responsibility, and methods are shown in Part I. The next step is to calculate the bearings and lengths between the stations, so that all station co-ordinates can be "proved" and any displaced stations re-instated. If possible, the site engineer should arrange for the engineer's representative to check his calculations and later to check readings and measurements taken on site. For the purpose of the example stations 8—13 are chosen as the base reference line.

Referring to Fig. 27, bearings from 8—13; co-ordinates

	E	N
Station 8	9353.936	3958.294
Station 13	9180.659	4020.936
Difference	173.277 (West)	62.642

$$\text{Tan } \alpha, \text{ quadrant bearing, (QB)} = \frac{\text{Diff E}}{\text{Diff N}} = \frac{173.277}{62.642}$$

$$\alpha = 70° \; 7'\cdot 28 \; (\text{N.W.})$$

Therefore, whole circle bearing (WCB) = 289° 52′ 32″

Distance 8—13 = $\sqrt{173.277^2 + 62.642^2}$ = 184.252 metres

bearings 8—10
co-ordinates

	E	N
Station 8	9353.936	3958.294
Station 10	9290.208	4152.296
	63.728 (West)	194.002

W.C.B. = 18° 11′ 6″ (NW)
= 341° 48′ 54″

Distance 8 to 10 = 204.201 metres

51

Similarly, bearings and distance 8 to 5 and check bearings 13 to 10 and 13 to 5 can be calculated. These stations can now be proved on site and any displaced stations reinstated. The angle between stations shown should coincide with an accuracy better than 15" of arc and distance should agree to better than 20mm. The engineer's representative should be informed if these amounts are exceeded. The site engineer, when checking and measuring distances on the ground, must make appropriate allowances for ground slope. This can be done **either** by measuring the vertical angle between the points and by trigonometry calculating the increase in measured length **or** by taking the difference in level between the two points and provided the slope is not steeper than 1:8, using a simple formula

$$\frac{H^2}{2 \times \text{horizontal (Grid) distance}}$$, where H is the difference in level.

Adjustments for slope measure may be required several times in any distance to be measured, particularly if there are hillocks and valleys along the line.

When the station peg positions have been "proved" on site, the site engineer can calculate the bearings and lengths of the main points and centre lines of roads 34 and 31 in the example. This is made by drawing sketch skeleton layouts as shown in Figs. 28 and 29. Figure 29 shows the centre lines and co-ordinates of centre points and intersection points. Tangent points are referenced, e.g. 34/1a, 34/1b, etc. Figure 30 is a basis for recording calculated bearings and distances for use on site.

The calculations should be done in the peace and quiet of an office; 7 figure tables of trigonometrical functions are a necessity and a small desk top calculating machine with an 8 digital display with overflow will make for speedier and more accurate results.

Starting calculation from Station 8 and fixing the centre point of the turning circle on road 34 (0/34), co-ordinates are computed into bearings and distances thus:

	E	N
Station 8	9353.936	3958.294
0/34	9372.500	3999.750
	18.564	41.456

$$QB = 24° 7' 22'' \text{ NE}$$

Therefore, W.C.B. $= 24° 7' 22''$

The distance (Grid) from 8 to 0/34 $= \sqrt{18.564^2 + 41.465^2} = 45.422\text{m}$

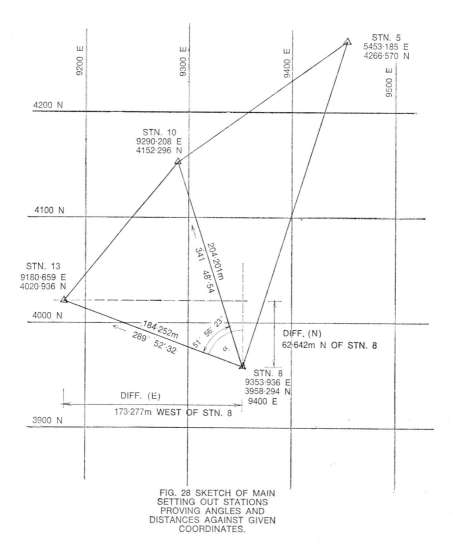

FIG. 28 SKETCH OF MAIN SETTING OUT STATIONS PROVING ANGLES AND DISTANCES AGAINST GIVEN COORDINATES.

These calculations are repeated along the main line of roads 34 and 31. To ensure the setting out "closes", check bearings and distances from alternative stations are added.

The site engineer can now start setting out the main points on site using colour coded pegs as described in Part I. Setting up the theodolite over station 8 fixing bearing 289 52' 32 sight on station 13. Then reading bearing 24 07' 22, measure 45.422m taking into account slope. This will determine point 0/34. This sequence is repeated along the main lines of roads 34 and 31

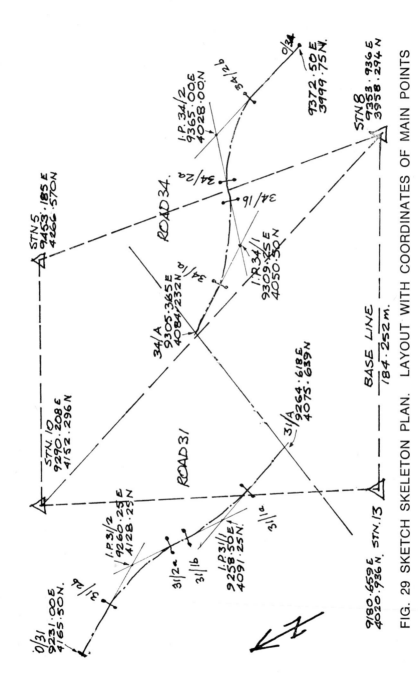

FIG. 29 SKETCH SKELETON PLAN. LAYOUT WITH COORDINATES OF MAIN POINTS

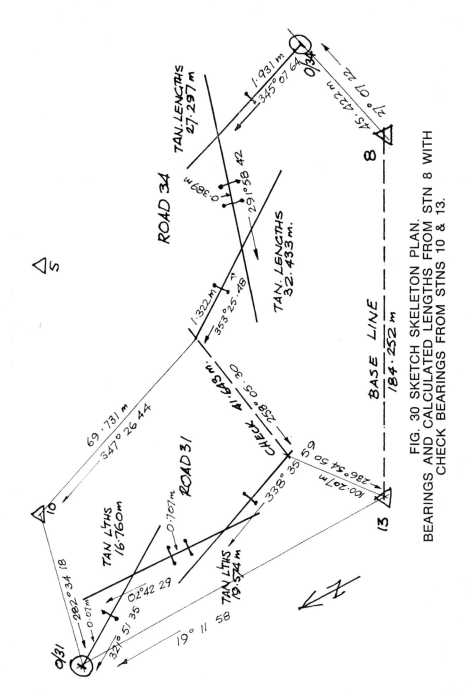

FIG. 30 SKETCH SKELETON PLAN.
BEARINGS AND CALCULATED LENGTHS FROM STN 8 WITH CHECK BEARINGS FROM STNS 10 & 13.

starting at 0/34 as shown in Fig. 30. Check bearings are taken from station 13 and station 10 to 0/31 and the setting out should close to within 50mm.

Note that in setting out main lines and points in roadworks no attempt is made at this stage in setting the road curves. The curve ranging should be done when main lines and points have been established and setting out closed and accuracy checked.

B. Vertical Curves

A worked example of a simple vertical curve is shown below. Calculations of this type should always be carried out on suitable pro-formas. Because of the amount of calculation required prior to setting and curves on site, it is vital that a proper record is kept of each step of the calculations.

The elements of a vertical curve and the necessary formulae are shown below in Fig. 31.

A worked example from a typical longitudinal section, together with completed pro-formas, is illustrated in Fig. 32.

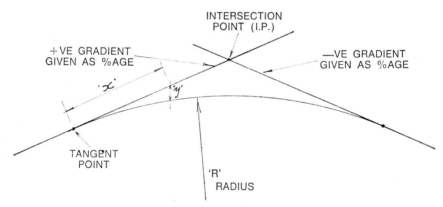

$$R = \frac{\text{Distance between Tan. Points} \times 100}{\text{Algebraic diff. in the two gradients}} \qquad y = \frac{x^2}{2R}$$

Where x is always the distance of the point chosen to the Tan. Point from which the Curve is to be set out.

NOTE: Gradients can both be +ve (or −ve) when a curve joins two slopes both going in the same general direction e.g.

FIG. 31 ELEMENTS OF A VERTICAL CURVE

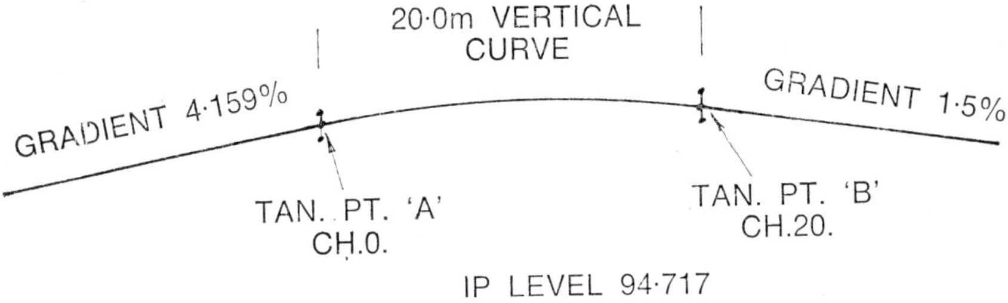

```
                    20·0m VERTICAL
                        CURVE
GRADIENT 4·159%                            GRADIENT 1·5%
           _____
                  TAN. PT. 'A'        TAN. PT. 'B'
                     CH.0.              CH.20.
```

IP LEVEL 94·717

Taken from Longitudinal
Section of a Road Layout

Level at Tan. Pt. A = I.P. Level $- \left(10\cdot0\text{m} \times \dfrac{4\cdot159}{100}\right)$

$= 94\cdot717 - 0\cdot416$
$= \underline{94\cdot301}$

Level at Tan. Pt. B $= 94\cdot717 - \left(10\cdot0\text{m} \times \dfrac{1\cdot5}{100}\right)$

$= \underline{94\cdot567}$

Level on gradient at CH.5
$= 94\cdot301 + \left(5 \times \dfrac{4\cdot159}{100}\right)$

$= \underline{94\cdot509}$

Level on ₵ of Curve at CH.5
$= 94\cdot509 - y$ (at CH.5)
$= 94\cdot509 - 0\cdot035$
$= \underline{94\cdot474}$

$R = \dfrac{20 \times 100}{(+4\cdot159) - (-1\cdot5)}$

$= \dfrac{2000}{5\cdot659}$

$= \underline{353\cdot419\text{m}}$

Levels are required at 5·00m intervals along curve from Tan. Pt. 'A'
Therefore $x = 5$

$y = \dfrac{5 \times 5}{2 \times 353\cdot419}$ at CH. 5·0m

$= \underline{0\cdot035\text{m}}$

CONTINUE CALCULATIONS AND ENTER ON PRO-FORMA

Pt.	REMARKS	CH. m	DIFF. IN CH. m	DIFF. IN LEVEL ON GRADE	RED. LEV. ON GRADE	$\dfrac{x^2}{2R}$	RED. LEV. ON ₵
tan. 'Pt. A	+4·159%	0	—	—	94·301	—	94·301
	Position of IP	5	5	0·208	94·509	−·035	94·474
		10	5	0·416	94·717*	·142	94·575
		15	5	0·075*	94·642	·035	94·607
tan. ·Pt. B	−1·5%	20	5	—	94·567	—	94·567

* USE LEV. AT IP REDUCE THIS BY 5 × 1·5%
TO ARRIVE AT RED. LEV. ON GRADE AT CH. 15

FIG. 32 WORKED EXAMPLE AND PRO FORMA
FOR SETTING OUT A VERTICAL CURVE

The basic procedure for calculating vertical curves is:
1. Calculate the levels on the vertical curve and fill in the pro-forma using the information shown on drawings or road gradients and length of curve. If the information is not

shown, then calculate the road gradient from the levels on the cross sections and decide on a suitable length of curve. Remember that this must be done in conjunction with, and the agreement of, the engineer's representative.
2. The intervals on the curve at which levels are required will depend on the degree of curvature and gradient. Normally, levels should be set at not more than 10m intervals.
3. In addition to (2) above, the chainage and level of the summit of the curve must be established. Where a "sag" occurs, the chainage and level of the lowest point must be found.

C. Horizontal Curves

The deflection angles for these curves are calculated from the formulae shown in the elements of a horizontal curve, Fig. 33.

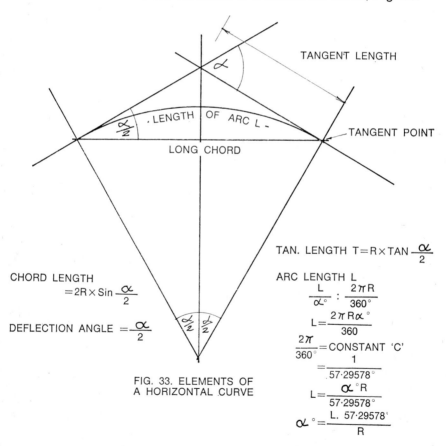

FIG. 33. ELEMENTS OF A HORIZONTAL CURVE

TANGENT LENGTH
LENGTH OF ARC L
LONG CHORD
TANGENT POINT

CHORD LENGTH
$= 2R \times \sin \frac{\alpha}{2}$

DEFLECTION ANGLE $= \frac{\alpha}{2}$

TAN. LENGTH $T = R \times \tan \frac{\alpha}{2}$

ARC LENGTH L
$\frac{L}{\alpha°} : \frac{2\pi R}{360°}$

$L = \frac{2\pi R \alpha°}{360}$

$\frac{2\pi}{360°} = $ CONSTANT 'C' $= \frac{1}{57 \cdot 29578°}$

$L = \frac{\alpha° R}{57 \cdot 29578°}$

$\alpha° = \frac{L. \; 57 \cdot 29578'}{R}$

A worked example based on the preceding estate roads layout is shown in Fig. 34. The deflection angles and bearings, together with chord lengths, should be calculated and entered onto a pro-forma as illustrated.

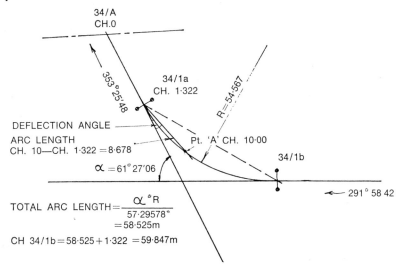

FIG. 34 SETTING OUT CENTRE LINE OF CURVE ON ESTATE ROAD LAYOUT (FIG. 29). ROAD NO. 34 FROM 34/A, 34/1a TO 34/1b (FROM FIG. 27 RADIUS=54·567m TAN LENGTHS=32·44m).

STATION (1)	CHAINAGE OF CHORD PT m (2)	ARC LENGTH m (3)	DEFLECTION ANGLE (4) DEG. MIN. SEC.			W.C.B. READING (5) DEG. MIN. SEC.			CHORD LENGTH m (6)
34/1a	1·322					173	25	48	
'A'	10·000	3·678	04	33	22	168	52	26	8·669
	20·000	10·000	05	15	00	163	37	26	9·986
	30·000	10·000	05	15	00	158	22	26	9·986
	40·000	10·000	05	15	00	153	07	26	9·986
	50·000	10·000	05	15	00	147	52	26	9·986
34/1b	59·847	9·847	05	10	11	142	42	15	9·834
		$\frac{\alpha}{2}=$ (30	43	33)					
		$\alpha=$ 61	27	06		(111	58	42)	

DEFLECTION ANGLE OF PT. 'A'

$$'\alpha' = \frac{L\ 57\cdot29578}{R}$$

$$\frac{57\cdot29578}{54\cdot567} = 1\cdot05001$$

$$\frac{'\alpha'}{2} = \frac{8\cdot678 \times 1\cdot05001}{2}$$

Def. Ang. $= 4° \ 35' \ 22''$

W.C.B. of Pt. 'A' CH 10·0m from 34/1a
   $= 353° \ 25' \ 48'' - 180° - 4° \ 33' \ 22''$
   $= 168° \ 52' \ 26''$

Chord Length from 34/1a to Pt. 'A'

   $= 2R \times \text{Sin } \alpha$

   $= 2 \times 54.567 \times \text{Sin } 4° \ 33' \ 22''$

   $= 8\cdot669\text{m}$

Pt. 'A' CAN NOW BE FIXED

Using proforma shown on previous page, calculate all the deflection angles at all the chosen points along the arc (col. 4). The sum of these deflections should equal $\frac{\alpha}{2}$ from Fig. 33.

The W.C.B. are then worked out to each of the chainages (Col. 5).

Assuming theodolite is set up at Tan. Pt. 34/1a. the final bearing to 34/1b $= 142° \ 42' \ 15''$

To check :—

   $142° \ 42' \ 15'' =$ The bearing of the Tan.straight
   IP 34/1 to IP 34/2 $= 291° \ 58' \ 42'' - 180° = 111° \ 58' \ 42''$ (see Fig. 30)

   Final bearing $= 142° \ 42' \ 15'' -$ sum of deflections
   $= 142° \ 42' \ 15'' - 30° \ 43' \ 33''$
   $= 111° \ 58' \ 42''$

Finally calculate chord lengths (col. 6).

The interval between points on the curve will depend upon a number of factors:
1. The degree of curvature, e.g., a sharp curve requires more chords.
2. The type of construction, e.g., excavation and earthworks intervals from 20—30m. Kerb and channel setting from 5—10m depending on curvature. Retaining walls and more complex structures from 2—5m.
3. The slope of the ground, e.g., when the level changes greatly, intervals must be set which will "catch" the change in level.

D. Spiral Transition Curves

The setting out of spiral transition curves is now greatly simplified following the publication of the Highway Transition Curve Tables compiled by the County Surveyors Society.

The elements of spiral transition curves are shown, Fig. 35:

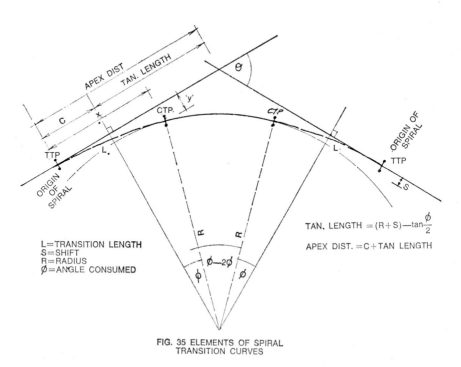

FIG. 35 ELEMENTS OF SPIRAL TRANSITION CURVES

It is usual for the engineer to specify the radius and the appropriate transition curve table to be used. It is also usual to supply the contractor with the deflection angles and chainages in the form of a computer print out. An example of such a print out is shown in Fig. 37, and a worked example is shown below, Fig. 36.

EXAMPLE

Given the radius and the appropriate table No. by the employing authority all curve data is obtained from Highway Transition Curve Tables (metric) compiled by the County Surveyors Society.

Using the computer printout Fig. 37 which is supplied by the employing authority (which in this case does not relate to the previously mentioned Curve Tables).

TTP is at CH. 812·388.

Coordinates; X 479523·902, Y 262516·781

Bearing 142° 6′ 50″

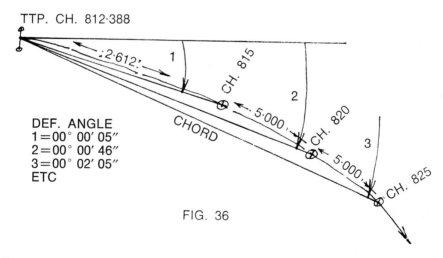

FIG. 36

RIGHT HAND TRANSITION CURVE
PRECEDING ELEMENT 4

TOTAL LENGTH=115·000		SHIFT=1·449		C=57·456		RL=VALUE =43700·000			
	CHAINAGE	x		y		BEARING OF TANGENT		CHORD	DEFLECTION ANGLE
ORIGIN	812·388	479523·902		262516·781		142 6 50			
	812·388	479523·902		262516·781		142 6 50		2·612	0 0 5
	815·000	479525·507		262514·719		142 7 6		5·000	0 0 46
	820·000	479528·576		262510·772		142 9 6		5·000	0 2 5
	825·000	479531·642		262506·822		142 13 5		5·000	0 4 4
	830·000	479534·702		262502·868		142 19 2		5·000	0 4 42
	835·000	479537·754		262498·908		142 26 56		5·000	0 6 0
	840·000	479540·795		262494·940		142 36 49		5·000	0 10 57
	845·000	479543·825		262490·962		142 48 40		5·000	0 13 33
	850·000	479546·839		262486·973		143 2 28		5·000	0 18 48
	855·000	479549·836		262482·970		143 18 15		5·000	0 23 43
	860·000	479552·814		262478·954		143 36 0		5·000	0 29 18
	865·000	479555·770		262474·921		143 55 42		5·000	0 36 31
	870·000	479558·701		262470·870		144 17 23		5·000	0 43 24
	875·000	479561·606		262466·801		144 41 2		5·000	0 51 56
	880·000	479564·481		262462·710		145 6 38		5·000	0 59 8
	885·000	479567·325		262458·598		145 34 13		5·000	1 9 58
	890·000	479570·134		262454·462		146 3 46		5·000	1 18 29
	895·000	479572·907		262450·301		146 35 16		5·000	1 29 38
	900·000	479575·640		262446·114		147 8 45		5·000	1 40 27
	905·000	479578·331		262441·900		147 44 12		5·000	1 52 55
	910·000	479580·977		262437·657		148 21 36		5·000	2 4 55

MOVE INSTRUMENT
LONG CHORD * BACK ANGLE TO STATION AT CHAINAGE   812·388   ARE   97·561   4 9 52

	915·000	479583·576		262433·386		149 0 59		5·000	0 19 32
	920·000	479586·124		262429·084		149 42 20		5·000	0 39 42
	925·000	479588·619		262424·751		150 25 38		5·000	1 0 32
	927·388	479589·791		262422·671		150 47 1		2·388	1 10 43

MOVE INSTRUMENT
LONG CHORD * BACK ANGLE TO STATION AT CHAINAGE   910·000   ARE   17·386   1 14 41

MOVE INSTRUMENT
LONG CHORD * BACK ANGLE TO STATION AT CHAINAGE   812·388   ARE   114·883   5 46 49

END OF TRANSITION
ACTUAL LENGTH=115·000

FIG 37 COMPUTER PRINTOUT FOR
WORKED EXAMPLE TRANSITION CURVE

63

It is necessary on occasions for the site engineer to calculate his own deflection angles. For instance, the print out may not show a position on the curve of a structure which may cross the highway. To calculate any deflection angle, for example the first deflection angle:

$$RL = 43700.00$$
$$L = CH\ 815 - CH\ 812.388 = 2.612$$

Therefore R @ $CH\ 815 = \dfrac{43700.000}{2.612} = 16730.4747$

Therefore $\emptyset$ (in Radians) $= \dfrac{L}{2R} = \dfrac{2.612}{33,460.9495}$

$$= .0000780611$$

The formula for an ideal spiral.

$$\text{Tan Def. Angle} = \emptyset + \dfrac{\emptyset^3}{3} + \dfrac{\emptyset^5}{105} + \dfrac{\emptyset^5}{5997} + \ldots\ldots\ldots$$

Note:
(For normal usage three terms are sufficient. If the spiral is very sharp, it may be necessary to use more terms.)

$$= 0.000026020$$
$$= 00°\ \ 00'\ \ \ \ 05''$$

To set out the deflection angles: set up the theodolite on the origin (TTP), define the line of tangent at bearing 142° 6′ 50″, knock in a peg giving the line. Sight along line and set instrument at 0° or 360°, then follow the deflections given until it is necessary to move the instrument. Move the theodolite to the last point on the curve previously set out. Set the instrument at 4° 9′ 52″ sighting back to origin (TTP) along the long chord, zero the instrument, then recommence reading deflection angles as given.

PART 4
Structures

I. INTRODUCTION

The setting out work involved in structures both above and below ground falls into two broad categories.
1. The location of the structure or group of structures on the site. This involves the establishment of the centre lines and corner positions of a building in relation to some site reference point on base line.
2. The location of the elements of the structure, relative to each other, within the boundaries of the structure. This covers the positioning of piles, columns, beams, etc., as construction proceeds.

This part of the manual describes procedures for both these types of setting out, as well as dealing with the special problems associated with certain types of structure, e.g., circular structures.

II. LOCATION OF STRUCTURES ON THE SITE

A. General Principles

An example of setting out roadworks using co-ordinates is shown in Part 3, Section V. The process is exactly the same for setting out structures. If a local grid has been provided by the engineer, then the referencing of setting out stations and the main features of the structures is greatly facilitated.

Increasing use is now made of National Grid Co-ordinates in major contracts, and an explanation of this system is given in Part 3, Section IIC. Alternatively, the contract drawings will indicate a base line from which the structures can be set out by triangulation. Should no particular system be indicated by the engineer, there is, therefore, a choice between a grid covering the site or a base line. In general, a grid is to be preferred on a larger site where a number of structures are to be related to each other. A base line is more suited for the smaller site and where obstacles will not obscure the use of instruments. Whichever method of setting out is chosen, the site engineer must agree this with the engineer's representative.

Before starting setting out on site, the engineer must decide, in the case of a grid referencing system, where he is going to place the main setting out stations and calculate bearings and distances. The bearings and distances from these stations to the chosen features of the structures can also be calculated.

These features can be centre lines and /or corner positions, and, in the case of curved structures, centre points, tangent points and/or intersection points.

Where a base line has been chosen, its bearing and length is calculated together with angles and distances to features of the structure. Offset reference pegs will also need to be positioned, so that each end of the base line can be re-established at any time during the works.

The site engineer should make a skeleton drawing of the outlines and centre lines of each structure, together with the stations and base line. These drawings should be recorded on site information sheets and copies kept in the site office. The provision of TBM's on the site should also be planned at this stage. In respect of this, it is vital to check with the engineer's representative which local Ordnance Bench Mark is to be used and to follow the procedures for TBM's described in Part I.

B. Initial Checks

Before starting setting out work on site, the site engineer must check the drawings in detail. Particular attention must be paid to the following points:—

1. See that the drawings are the latest editions supplied by the engineer's representative.
2. Sum all intermediate dimensions and ensure that the totals reconcile with overall dimensions.
3. Verify levels — levels should be clearly shown and reconciled with figured dimensions.
4. Verify the tolerance to be used in setting out the structure.
5. Similarly, check the positions and fixings for steelworks, pipework, etc.
6. Check reinforcement details with bending schedules. Although this may not have a direct bearing on early setting out work, the sooner any mistakes are found the less likely a delay to the contract will result.
7. Dimensions marked on drawings should always be given preference to those obtained by scaling. Scaling is bad practice and should be avoided where possible.

The most important check after the above items will involve work on site. That is "proving" the site drawings. In principle, proving will follow the process described in Part 3 and involves checking that the structures will fit into the site in the position indicated by the drawing. Proving will involve the setting up of some referencing system, if one is not already provided. It is convenient at the same time to set up the permanent stations and TBM's.

If, as a result of the various checking procedures, errors or inconsistencies are discovered, they must be referred to the engineer's representative so that he can take early action.

C. Relating One Structure To Another On Site

A project may involve several structures on the same site. The site engineer must ensure that individual units relate to each other, so that the whole site is laid out in accordance with the drawings. This is particularly important when pipeworks or services connect one structure with another structure. It is vital that the structures are set out to the tolerances specified for each structure. If pipework connects two structures, the site engineer must not take the degree of tolerance needed for the pipework as the relative tolerance in positioning the structures.

A very large site with a number of related structures poses additional problems which are mainly organisational. The site engineer in this case may find himself one of a team, each engineer being responsible for the setting out and control of a structure or part of a structure.

Under these circumstances, it is difficult to co-ordinate each part of the total setting out activity. In this situation, there is no substitute for a grid of the whole site with a series of reference

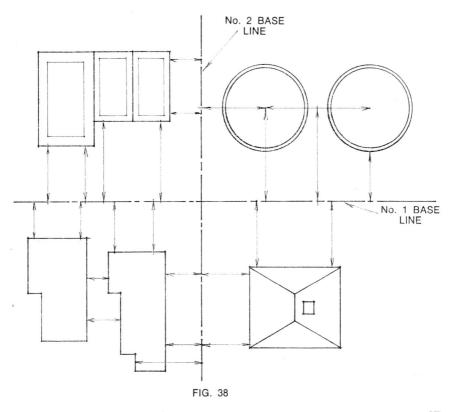

FIG. 38

stations established round the site so that every part of a structure can be defined by co-ordinates and bearings to the stations.

For a small site a simple system of base lines and offsets can be used as shown in Fig. 38. This is a diagram of sewage disposal works using two base lines at right angles.

To ensure accuracy of setting out on sloping sites, corrections should be made to plan measurements using the formula $\dfrac{H^2}{2 \times \text{horizontal (Grid) distance}}$ where H is the difference in level.

III. SETTING OUT AND CONTROL OF THE ELEMENTS OF THE STRUCTURE

A. Excavation

The size of the excavation will normally be decided by site management and overall dimensions will be given to the site engineer. The site engineer must take into account the space required round the excavation for access and plant utilisation. If possible, a standard offset dimension should be used for the whole job. The offset dimension should be a multiple of 1·0m. A sketch plan and section should be prepared on an information sheet and given to the site supervisor.

The excavation will normally involve two operations. The first will be the removal of topsoil followed by the actual excavation. For the removal of topsoil corner pegs only should be set. The actual excavation and the lines of the structure can then be defined by corner profiles as shown in Fig. 39.

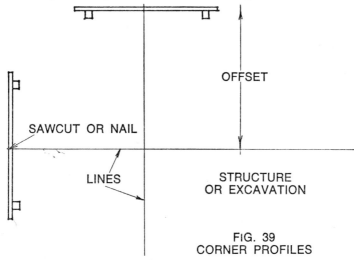

FIG. 39
CORNER PROFILES

When setting out large and complex structures with many corners and re-entrants, it is useful to paint the profiles on the vertical plan axis in different colours from the horizontal axis, e.g., vertical in red and white, horizontal, black and white.

The lines can be set up from the profiles, and before excavation commences a thin line of sand or lime should be sprinkled on the ground before removing the lines to give a guide to the excavation machinery. A point to remember is that grass and low vegetation provide a good protective cover to the ground. Over-removal of topsoil tends to make transport about the site that much more difficult. Profiles should be constructed, so that the top surface of the rail is to a reduced level. Levels should be transferred from level pegs close to each pair of profiles.

A traveller of appropriate length can then be used to control the depth of dig. The length of a traveller should not normally be more than 5·00m. For deep excavation, it will be necessary to transfer levels in stages as excavation proceeds. Methods of doing this are described in Part 2, Section VIII.

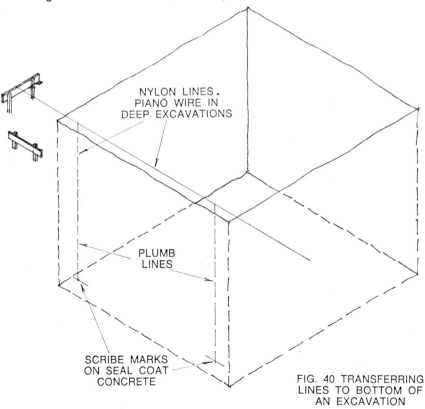

FIG. 40 TRANSFERRING LINES TO BOTTOM OF AN EXCAVATION

The dimensions of the structure can be transferred to the bottom of the excavation by use of a theodolite provided sufficient of the formation can be seen from ground level. This is not often the case with deep excavation or where the timbering impedes the sight lines. Nylon lines should then be set from profile to profile and the line transferred by means of a plumb line. The sag of the line is a problem in deep excavations which makes the stabilising of the plumb bob difficult.

In such situations piano wire is used in conjunction with a heavy plumb bob which should be damped in a bucket of water or oil.

The accurate transfer of structure lines is made easier where there is a binding layer of concrete. A black painted patch is dabbed on the concrete and the intersection point of two lines scratched in the drying paint with a scriber, Fig. 40.

The transfer of reduced levels to the bottom of the excavation is best done with a steel tape once the excavation is deeper than a normal staff. It is important to seek the confirmation and approval of the final base level by the engineer's representative once it is established.

B. Piling

The main problems in setting out and controlling piling work, whether driven or bored piles, is the disturbance of the ground and the amount of room required by the piling rigs.

Piles are normally driven to tolerance of + or − 75mm in verticality in 20m of length. The position for the pile at ground level must be precise and the piling tolerance must never be taken into account when setting out positions for piles.

It is common practice for the contractor to sublet the piling content of the job to a specialist piling sub-contractor. The site engineer must familiarise himself with the terms of the sub-contract agreement. Some specialists prefer to set out their own pile positions with their own engineer. Others put the responsibility for setting out the piles with the main contractor. Whichever terms apply to a particular job the final responsibility for correct setting out remains with the main contractor.

Because of the general disruption caused by piling rigs, it is usually impossible to set out pile positions much in advance of the work without constant re-location of pins. This is particularly so with large augured piles where there is a considerable amount of soil to remove from the boring operation. The engineer must therefore be in close attendance on the rigs so that he can set out the positions just in front of them.

A suitable method the site engineer can use to ensure the quick and accurate positioning of piles just ahead of the rigs is shown in Fig. 41.

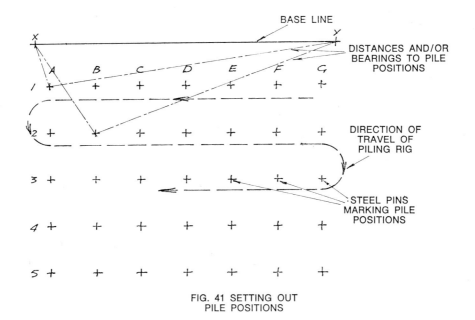

FIG. 41 SETTING OUT PILE POSITIONS

Before piling work commences, the pins are set to the required pile positions from the corner profiles using a bricklayer's line and a steel tape. A base line is defined and a steel tape is used to measure the distances to each pin. The pins and lines are then removed. A suitable table of pile references can then be drawn up and recorded on a site information sheet. The base line should be in such a position that taped measurements can be taken unimpeded by the rig. This method is convenient only when the maximum distances do not exceed a steel tape length — say 60m maximum, or when the number of piles is not great.

When distances are greater than 60m, it is better to use a theodolite and calculate angles and distances from the base line to each pile position. The site engineer must remember that, apart from the problem of disruption by pins or rigs, the movement of the rig and the displacement of substrata by the pile causes a "wave" of ground in front of the rig. Pile positions can therefore be displaced by a considerable amount from the true position.

If ground conditions are reasonable, there will be a number of bores drilled before any concrete is poured. As soon as a bore is clear, and the rig moved on, a temporary level peg should be set near the bore so that the depth can be checked using a weighted steel tape and the top level for concrete defined.

C. Stanchion Bases

Stanchion bases should be set out using the centre lines of the stanchions. Holding down bolts or formed mortices must be positioned using a plywood template on which the centre lines are marked and which is drilled for the bolts. The template and bolts are fixed in position with pegs and struts suspended over the excavation. The concrete must be placed with care, and at each stanchion base the site engineer should:

1. See that the template is not displaced by the concrete placing.
2. Check the position directly concrete has been placed.
3. Check the position again before the final set of the concrete.

A temporary level peg should be provided for each stanchion base, so that the finished level of the top of the base can be controlled.

D. Reinforced Concrete Columns and Walls

The site engineer must provide the line and level of the kickers for walls and columns. The careful control of the construction of the kicker formwork will pay dividends when fixing the formwork to the walls and columns later. The wall and column formwork should be controlled for verticality by making offsets on the slab as shown in Fig. 42.

The marks should be scribed with a steel point on a black paint patch. It is best to adhere to a standard offset dimension for the job — 0·5m is usually most practical. The line of the formwork must be checked by the site engineer before and during placement of concrete. This can be done with a bricklayer's line or a theodolite.

Temporary fixing blocks for services and the like in concrete should be positioned and checked in the same way as for holding bolts described above. The site engineer should sketch the details of holes for pipe or services, properly dimensioned, on a site information sheet and give copies to the general foreman and foreman carpenter.

Control of the top level of concrete in wall formwork is best done by fixing battens to the faces of the forms, with the bottom edge of the batten corresponding to the finished level of the concrete. The site engineer should mark the position for the battens by establishing reduced levels at 3·0m intervals and fixing a row of small nails between the levels.

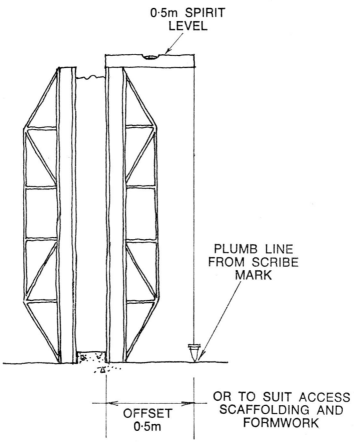

FIG. 42 CONTROL OF
VERTICALITY OF FORMWORK
FOR WALLS & COLUMNS

IV. MULTI-STOREY STRUCTURES

A. Verticality

The maintenance and control of verticality is the most important setting out feature of multi-storey structures. Most of these structures involve the installation of lifts. As the tolerance in vertical working of lift equipment is very small, it is usual to position reference points on each of the lift well walls. Initially, these points should be set in the lift well base.

The control of verticality is done by using a heavy plumb bob suspended on piano wire. Reference lines for the structure can then be projected through the lift well reference points, and internal walls and columns measured from the lines.

The use of any plumb bob becomes difficult and time consuming when the height of a structure exceeds about 20m. The projection of plumb marks as reference lines can also lead to inaccuracies over large floor areas. The provision of vertical guides to structures over 20m height and/or of large plan area should be done either by:
1. Using a theodolite sited well away from the structure and using the vertical curve to measure vertical angles and distances.
2. Using an optical plumbing device within the confines of the structure.

The problem with the first method is that the instrument must be sited at a distance at least twice the vertical height of the building for accuracy. This becomes difficult in urban areas where the sight line from the instrument is impeded by existing structures.

The second method involves control floor by floor by siting through perspex or similar small windows cast in the floors. The instrument is set up over the "window" sighting down onto the reference points in the base. The sight is then made upwards onto a target from which control dimensions can be taken for column and wall forms.

These instruments have many advantages, including both speed and accuracy. It is also easy to provide a number of plumbing points. These will usually be positioned at each corner of the structure and at lift wells, stair wells, service ducts etc.

B. Height and Level

The control of floor-to-floor dimensions is usually done by a weighted steel tape, measuring each time from a datum in the base of the structure. Each floor is then provided with datum marks in key positions from which a quick set level can be used to transfer levels on each floor. The base datum levels should be set in the bottom of lift wells, service ducts, window mullions and any location which provides an unrestricted taping line to roof level.

V. CURVED STRUCTURES

Wherever the site engineer can locate the centre point, and the radius is within the length of steel tape available, he should use the centre to set out circular curves.

The methods in Part 4 for setting out circular curves for roadworks apply to circular curved structures, where the centre point cannot be conveniently used for striking the curves.

The curve must then be set out using deflection angles and chord lengths as described. The length of chord will be determined by the degree of curvature, and, more important, the type of wall construction. If of reinforced concrete, the chord intervals will be governed by the size of the formwork panel.

The centre line of the wall is first pegged out, offset pegs established, profiles erected and the foundation dug and concreted.

The centre line must then be re-established. Offset scribe marks are then made at about 0·5m away from centre. The kicker is constructed with great care both to curve and level. The main wall forms are then erected and verticality maintained by offset plumbing down to the scribe marks.

Construction of masonry or brick curved walls should follow the same principles as reinforced concrete. The site engineer should provide chord intervals of about 3·0m. A timber curved former can then be used cut to the profile of the curve. The bricks are then laid course by course to the shape of the former. The site engineer should assist the tradesmen to ensure that the first course of bricks follows a smooth curve round its whole length. The accuracy of this first course of brick will determine the accuracy of the whole wall.

Structures which in plan consist of a series of sections from circular curves, for example, an 'S' shaped building, are difficult to set out. This difficulty is not one of principle, but one of monotonous calculation of angles and offsets while at the same time maintaining sufficient accuracy.

The use of a computer to provide chord lengths and deflection angles for chord structures is to be recommended.

VI. CIRCULAR STRUCTURES

The most usual circular structure the site engineer will be required to set out is some form of concrete or brickwork tank. Digestion tanks in a sewage treatment works are an example of this, Fig. 43.

The centre point of the inverted cone form of the structure would first be defined and four reference pegs set, so that the centre point can always be re-established. Excavation is then completed and the centre base concrete laid. The exact centre point is then re-established from the reference pegs.

A tripod and platform is then erected and the drilled hole in the platform plumbed to the base centre point. A bolt is then fixed in the hole in the platform. This bolt can then be used as a radius point from which a steel tape can be used at the various stages of construction of the sloping floors. Exact measurements

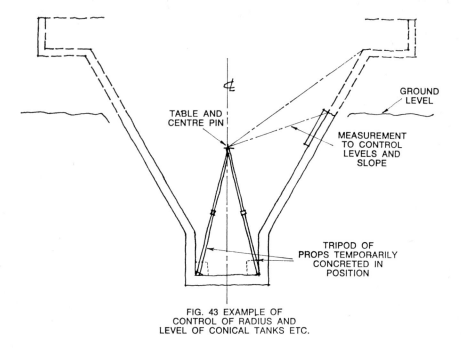

FIG. 43 EXAMPLE OF CONTROL OF RADIUS AND LEVEL OF CONICAL TANKS ETC.

should be calculated by the site engineer to the blinding concrete surface, reinforcement position and top formwork face for each lift of formwork.

The above is just one example where the site engineer will need his ingenuity to solve a tricky site control problem in such a way that structures can be maintained within the tolerances specified.

The curve must then be set out using deflection angles and chord lengths as described. The length of chord will be determined by the degree of curvature, and, more important, the type of wall construction. If of reinforced concrete, the chord intervals will be governed by the size of the formwork panel.

The centre line of the wall is first pegged out, offset pegs established, profiles erected and the foundation dug and concreted.

The centre line must then be re-established. Offset scribe marks are then made at about 0·5m away from centre. The kicker is constructed with great care both to curve and level. The main wall forms are then erected and verticality maintained by offset plumbing down to the scribe marks.

Construction of masonry or brick curved walls should follow the same principles as reinforced concrete. The site engineer should provide chord intervals of about 3·0m. A timber curved former can then be used cut to the profile of the curve. The bricks are then laid course by course to the shape of the former. The site engineer should assist the tradesmen to ensure that the first course of bricks follows a smooth curve round its whole length. The accuracy of this first course of brick will determine the accuracy of the whole wall.

Structures which in plan consist of a series of sections from circular curves, for example, an 'S' shaped building, are difficult to set out. This difficulty is not one of principle, but one of monotonous calculation of angles and offsets while at the same time maintaining sufficient accuracy.

The use of a computer to provide chord lengths and deflection angles for chord structures is to be recommended.

VI. CIRCULAR STRUCTURES

The most usual circular structure the site engineer will be required to set out is some form of concrete or brickwork tank. Digestion tanks in a sewage treatment works are an example of this, Fig. 43.

The centre point of the inverted cone form of the structure would first be defined and four reference pegs set, so that the centre point can always be re-established. Excavation is then completed and the centre base concrete laid. The exact centre point is then re-established from the reference pegs.

A tripod and platform is then erected and the drilled hole in the platform plumbed to the base centre point. A bolt is then fixed in the hole in the platform. This bolt can then be used as a radius point from which a steel tape can be used at the various stages of construction of the sloping floors. Exact measurements

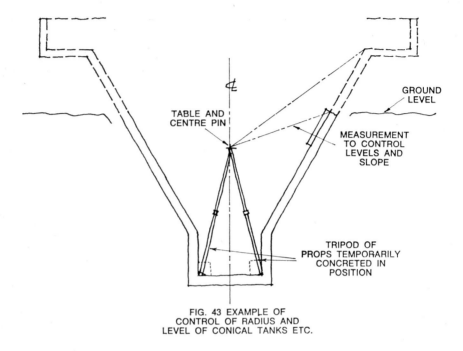

FIG. 43 EXAMPLE OF CONTROL OF RADIUS AND LEVEL OF CONICAL TANKS ETC.

should be calculated by the site engineer to the blinding concrete surface, reinforcement position and top formwork face for each lift of formwork.

The above is just one example where the site engineer will need his ingenuity to solve a tricky site control problem in such a way that structures can be maintained within the tolerances specified.